CONGRESSIONAL POLICIES, PRACTICES AND PROCEDURES

ADVANCED FOSSIL ENERGY, LICENSING OF PATENTED LABORATORY INVENTIONS AND CONSIDERATIONS FOR MAINTAINING US COMPETITIVENESS IN POTENTIALLY TRANSFORMATIONAL RESEARCH AREAS

Congressional Policies, Practices and Procedures

Additional books and e-books in this series can be found on Nova's website under the Series tab.

CONGRESSIONAL POLICIES, PRACTICES AND PROCEDURES

ADVANCED FOSSIL ENERGY, LICENSING OF PATENTED LABORATORY INVENTIONS AND CONSIDERATIONS FOR MAINTAINING US COMPETITIVENESS IN POTENTIALLY TRANSFORMATIONAL RESEARCH AREAS

EDWARD CARR
EDITOR

snova
New York

Library of Congress Cataloging-in-Publication Data

ISBN: 978-1-53615-071-1

Published by Nova Science Publishers, Inc. † New York

Contents

PREFACE

The government documents included in this book are comprised of reports and testimonies from June 2018 to September 2018 on research and development in the United States. The first report examines research and development projects started from 2010 through 2017 on advanced fossil energy. For more than 100 years, three fossil fuel sources – coal, oil and natural gas – have made up at least 80% of total US energy consumption. While fossil fuels are associated with some negative environmental impacts – such as carbon dioxide emissions, the predominance of coal, oil and natural gas is likely to continue into the future.

The second report focuses on considerations for maintaining US competitiveness in Quantum Computing, Synthetic Biology and other potentially transformational research advances. Federal support in such areas can accelerate innovation and drive technological advances and promote US competitiveness in the global economy.

The third and final report focuses on additional actions needed to improve licensing of patented laboratory inventions. The GAO was asked to review agency practices for managing inventions developed at federal labs, with a particular focus on patent licensing. This report examines the challenges in licensing patents and steps take to address and report them.

Chapter 1 - One aspect of DOE's mission is to secure U.S. leadership in energy technologies. To that end, DOE funds R&D for energy projects, including for advanced fossil energy (innovative technologies for coal, natural gas, and oil). DOE provides funding for R&D projects, including large projects designed to demonstrate the commercial viability of technologies. Also, DOE is authorized to make loan guarantees to support certain energy projects through its Loan Guarantee Program, which is administered by its Loan Programs Office.

GAO was asked to review DOE's funding for advanced fossil energy projects. This report describes DOE's funding for advanced fossil energy R&D projects started from fiscal years 2010 through 2017 and the types of projects and recipients that received funding, among other objectives. For purposes of this report, GAO used the term funding to mean obligations.

GAO analyzed relevant laws, regulations, and guidance; DOE data on R&D funding for fiscal years 2010 through 2017; and DOE documents. GAO also interviewed DOE officials in the Office of Fossil Energy, the National Energy Technology Laboratory, and the Loan Programs Office.

Chapter 2 - Scientific and technological innovation contributes to U.S. economic competitiveness and prosperity. Federal agencies support transformational technological advances—those that result in new or significantly enhanced technologies— by, for example, funding research (nearly $70 billion in obligations in fiscal year 2017).

GAO was asked to examine support for research that could lead to transformational technological advances. This report (1) describes federal agencies' and nonfederal entities' support for such research in selected areas, (2) examines federal agencies' coordination on this research, and (3) describes experts' views on considerations for maintaining U.S. competitiveness through such advances. GAO selected quantum computing and synthetic biology as examples of research areas that could lead to transformational technological advances. GAO reviewed agency documents and interviewed federal officials, subject matter experts, and stakeholders. GAO also worked with the National Academies of Sciences, Engineering, and Medicine to convene a meeting to solicit views from 19 experts selected from government, academia, and industry, among others.

Chapter 3 - The federal government spends approximately $137 billion annually on research and development—mostly at DOD, DOE, NASA, and NIH—to further agencies' missions, including at federal labs. Multiple laws have directed agencies and labs to encourage commercial use of their inventions, in part by licensing patents, to private sector companies and others that aim to further develop and bring the inventions to market.

GAO was asked to review agency practices for managing inventions developed at federal labs, with a particular focus on patent licensing. This report examines (1) challenges in licensing patents and steps taken to address and report them and (2) information to guide establishing financial terms in patent licenses at DOD, DOE, NASA, and NIH. GAO reviewed relevant literature, laws, and agency documents, including patent licenses from 2014, to match the most recent NIST summary report when the licenses were requested, and GAO interviewed agency officials and knowledgeable stakeholders, including organizations that assist federal labs in licensing patents.

In: Advanced Fossil Energy …
Editor: Edward Carr
ISBN: 978-1-53615-071-1

Chapter 1

ADVANCED FOSSIL ENERGY: INFORMATION ON DOE-PROVIDED FUNDING FOR RESEARCH AND DEVELOPMENT PROJECTS STARTED FROM FISCAL YEARS 2010 THROUGH 2017*

United States Government Accountability Office

WHY GAO DID THIS STUDY

One aspect of DOE's mission is to secure U.S. leadership in energy technologies. To that end, DOE funds R&D for energy projects, including for advanced fossil energy (innovative technologies for coal, natural gas, and oil). DOE provides funding for R&D projects, including large projects designed to demonstrate the commercial viability of technologies. Also,

* This is an edited, reformatted and augmented version of United States Government Accountability Office; Report to the Ranking Member, Subcommittee on Energy, Committee on Energy and Natural Resources, U.S. Senate, Publication No. GAO-18-619, dated September 2018.

DOE is authorized to make loan guarantees to support certain energy projects through its Loan Guarantee Program, which is administered by its Loan Programs Office.

GAO was asked to review DOE's funding for advanced fossil energy projects. This report describes DOE's funding for advanced fossil energy R&D projects started from fiscal years 2010 through 2017 and the types of projects and recipients that received funding, among other objectives. For purposes of this report, GAO used the term funding to mean obligations.

GAO analyzed relevant laws, regulations, and guidance; DOE data on R&D funding for fiscal years 2010 through 2017; and DOE documents. GAO also interviewed DOE officials in the Office of Fossil Energy, the National Energy Technology Laboratory, and the Loan Programs Office.

What GAO Recommends

GAO is not making any recommendations.

What GAO Found

The Department of Energy (DOE) provided $2.66 billion in funding, or obligations, for 794 research and development (R&D) projects started from fiscal years 2010 through 2017 to develop advanced fossil energy technologies. Such technologies include processes for converting coal into synthesis gas composed primarily of carbon monoxide and hydrogen, and recovering methane from gas hydrates. Of the $2.66 billion, DOE provided $1.12 billion in funding for 9 later-stage, large demonstration projects, which were to assess the readiness for commercial viability of carbon capture and storage (CCS) technologies. CCS involves capturing man-made carbon dioxide at its source and storing it permanently underground.

DOE provided the remaining $1.54 billion in funding for 785 other projects in amounts that were relatively small—over half were for less than $1 million.

Six demonstration projects researched CCS technologies using coal, while three used other fuels. The nine demonstration projects received funding ranging from $13 million to $284 million. As shown in the figure, three projects implementing CCS technologies were active as of the end of fiscal year 2017. Also, DOE withdrew its support for four projects, and two projects were withdrawn by the recipients—all before completion. These projects did not reach completion due to several factors, such as a lack of technical progress, or changes in the relative prices of coal and natural gas that made the projects economically unviable.

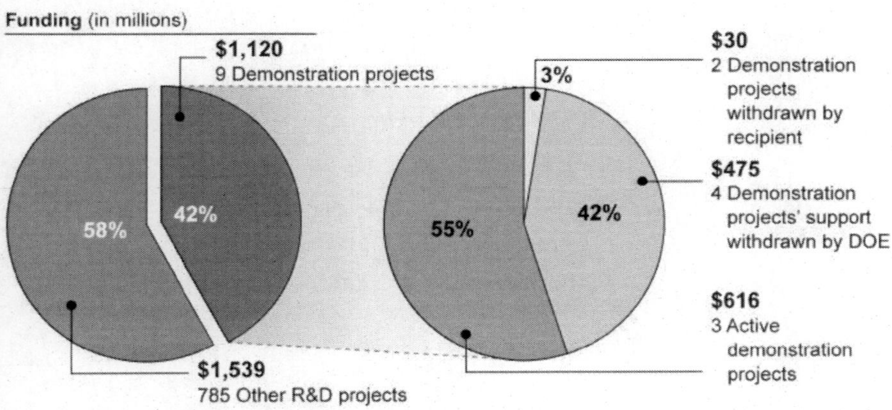

Source: GAO analysis of Department of Energy (DOE) data. | GAO-18-619.

Note: Of the 794 R&D projects, 9 large demonstration projects supported carbon capture and storage technologies to assess their readiness for commercial viability. Advanced fossil energy projects include research on innovative coal, natural gas, and oil technologies. For purposes of this report, GAO used the term funding to mean obligations. Funding amounts may not sum due to rounding.

Distribution of Department of Energy (DOE) Funding for Advanced Fossil Energy Research and Development (R&D) Projects that Started from Fiscal Years 2010 through 2017.

Of the 785 other projects, about 89 percent involved R&D of coal technologies, such as coal gasification—the conversion of carbon-containing material into synthesis gas. The other 11 percent of the 785 projects involved R&D of oil and gas technologies, such as the development of technologies to find, characterize, and recover methane from gas hydrates.

ABBREVIATIONS

CCS	carbon capture and storage
CCUS	carbon capture, utilization, and storage
CO_2	carbon dioxide
DOE	Department of Energy
EPAct	Energy Policy Act of 2005
FE	Office of Fossil Energy
IGCC	integrated gasification combined cycle
LGP	Loan Guarantee Program
NETL	National Energy Technology Laboratory
R&D	research and development
Recovery Act	American Recovery and Reinvestment Act of 2009

September 21, 2018

The Honorable Joe Manchin III
Ranking Member
Subcommittee on Energy
Committee on Energy and Natural Resources
United States Senate

Dear Senator Manchin:

For more than 100 years, three fossil fuel sources—coal, oil, and natural gas—have made up at least 80 percent of total U.S. energy consumption, according to the Department of Energy's (DOE) Energy Information Administration.[1] Furthermore, in 2017, about 63 percent of U.S. electricity generation was from fossil fuels—including coal, oil, and natural gas— and fossil fuels made up about 95 percent of transportation fuels in the United States.[2] While fossil fuels are associated with some negative environmental impacts—such as carbon dioxide (CO2) emissions—the predominance of coal, oil, and natural gas is likely to continue into the future, according to the Energy Information Administration.

Since 1979, DOE has provided technical and financial assistance for research and development (R&D) for advanced fossil energy projects. Advanced fossil energy projects include processes for capturing manmade CO2 at its source and storing it underground, converting carbon-containing material (i.e., coal) into synthesis gas composed primarily of carbon monoxide and hydrogen, and recovering methane from gas hydrates.[3] Within DOE, the Office of Fossil Energy (FE) carries out federal R&D programs on advanced fossil energy technologies, among other things. FE's primary mission is to ensure reliable fossil energy resources for clean, secure, and affordable energy while enhancing environmental protection. In addition, the National Energy Technology Laboratory (NETL)—overseen by FE—implements a broad spectrum of energy and environmental R&D programs; has expertise in coal, oil, and natural gas; and performs contract and project management, among other things. In addition, the laboratory coordinates and oversees R&D projects for FE.

---

[1] The Energy Information Administration is the statistical and analytical agency within DOE that collects, analyzes, and disseminates independent and impartial energy information.

[2] For transportation fuels, 92 percent were petroleum products—including gasoline, distillate fuels (mostly diesel fuel), and jet fuel—while 3 percent were from natural gas pipeline compressors.

[3] Synthesis gas is composed primarily of carbon monoxide and hydrogen, and can be used as a fuel to generate electricity or steam; as a basic chemical building block for a large number of uses in the petrochemical and refining industries; and for the production of hydrogen. Gas hydrates are naturally-occurring combinations of methane gas and water that form under specific conditions of low temperature and high pressure.

In addition to its R&D programs, DOE administers the Title XVII Loan Guarantee Program (LGP) for innovative energy technologies through its Loan Programs Office. One of the goals of the LGP is to encourage commercial use in the United States of new or significantly improved energy-related technologies. Title XVII of the Energy Policy Act of 2005 (EPAct), as amended, authorizes DOE to provide loan guarantees for certain energy projects in several categories, including advanced fossil energy technology.[4] Under the LGP, to encourage such innovative energy projects, the federal government agrees to reimburse lenders for the guaranteed amount of loans if the borrowers default. If borrowers default on the loans, the federal government can be exposed to substantial financial risks.

You asked us to review DOE's funding for advanced fossil energy projects. This report describes 1) DOE's funding for advanced fossil energy R&D projects that started from fiscal years 2010 through 2017 and the types of projects and recipients that received funding, and 2) DOE's loan guarantees, if any, for advanced fossil energy projects from fiscal year 2006 through August 2018. For the purposes of this report, we use the term funding to mean obligations.[5]

To address the first objective, we reviewed relevant laws, regulations, and DOE guidance. We analyzed DOE advanced fossil energy R&D project data for fiscal years 2010 through 2017. We focused our review on 794 advanced fossil energy R&D projects that received funding through NETL because the 794 projects represent all of the advanced fossil energy R&D projects in our scope started from fiscal years 2010 through 2017. We used fiscal year 2010 as the start date because DOE officials told us

---

[4] Pub. L. No. 109-58, Title XVII, 119 Stat. 594, 1117 (2005) (codified as amended at 42 U.S.C. §§ 16511-16516).

[5] An obligation is a definite commitment that creates a legal liability of the government for the payment of goods and services ordered or received, or a legal duty on the part of the United States that could mature into a legal liability by virtue of actions on the part of the other party beyond the control of the United States. Payment may be made immediately or in the future. An agency incurs an obligation, for example, when it places an order, signs a contract, awards a grant, purchases a service, or takes other actions that require the government to make payments to the public or from one government account to another. GAO, *A Glossary of Terms Used in the Federal Budget Process*, GAO-05-734SP (Washington, D.C.: Sept. 2005).

that DOE's current data management system came into use for the R&D projects that started in fiscal year 2010. We used fiscal year 2017 as the end date because that was the most recent complete year for which data were available. DOE provided us with a spreadsheet that included key project information—such as the name of the recipient of the R&D funding and the project start date—as well as obligations data for each project started during the period of our review by summing the obligations for the project from each year. We reported on DOE's funding for these R&D projects; DOE generally provided financial assistance for these projects through grants or cooperative agreements.[6] Because advanced fossil energy R&D financial assistance for the Small Business Innovation Research and Small Business Technology Transfer programs, as well as for the DOE national laboratories, are not obligated through NETL, we excluded these projects from our scope.[7] To assess the reliability of the funding data, as well as the specific project information for the 794 R&D projects, we interviewed data specialists at DOE Headquarters, FE, and NETL and reviewed DOE internal guidance for the maintenance of agency data. We found the data to be sufficiently reliable for our purposes. We also reviewed DOE websites and documentation, including fact sheets, and interviewed officials from FE and NETL. To characterize the recipients of advanced fossil energy R&D funding, we coded each recipient into one of

[6] Federal financial assistance means assistance that non-federal entities receive or administer in the form of grants, property, cooperative agreements, food commodities, direct appropriations, or other assistance, and can also include loans, loan guarantees, interest subsidies, and insurance, depending on the context, but does not include amounts received as reimbursement for services rendered to individuals in accordance with OMB-issued guidance. 2 C.F.R. § 200.40. See also 31 U.S.C. § 7501(5). A grant agreement is generally defined as a legal instrument of financial assistance between a federal awarding agency and a non-federal entity that is used to enter into a relationship the principal purpose of which is to transfer anything of value from the federal awarding agency to the non-federal entity to carry out a public purpose authorized by law, and not to acquire property or services for the federal awarding agency's direct benefit or use. 2 C.F.R. § 200.51. A cooperative agreement is distinguished from a grant in that it provides for substantial involvement between the federal awarding agency and the non-federal entity in carrying out the activity contemplated by the federal award. 2 C.F.R. § 200.24. For purposes of our report, we use the term awards to refer to both grants and cooperative agreements.

[7] We also did not include tax expenditures administered by the Department of the Treasury, such as the tax credits for industrial carbon dioxide capture and sequestration and investment in clean coal facilities.

three categories: university, industry or other.[8] We developed definitions for making the coding determination for each recipient; more information on this process, including the full definitions, is in appendix I.

To describe the status of DOE's advanced fossil energy loan guarantees, we reviewed relevant laws, regulations, and guidance, as well as past GAO reports describing DOE's administration of the loan program. We analyzed information that DOE provided on applications for loan guarantees for advanced fossil energy projects under the LGP and other related information for fiscal year 2006 through August 2018.

We used fiscal year 2006 as the start date because it was the first year that DOE issued an advanced fossil energy project solicitation—an announcement of opportunities for loan guarantees for advanced fossil energy projects— and we used August 2018 as the end date in order to provide the most up-to-date information possible. We also reviewed the advanced fossil energy project solicitations issued by DOE during this timeframe. To assess the reliability of the summary information, we interviewed LGP staff who maintain the information for advanced fossil energy applications, and reviewed DOE documentation. We found the data to be sufficiently reliable for our purposes. In addition, we interviewed officials from the Loan Programs Office who work on the LGP. For further details on our scope and methodology, see appendix I.

We conducted this performance audit from March 2017 to September 2018 in accordance with generally accepted government auditing standards. Those standards require that we plan and perform the audit to obtain sufficient, appropriate evidence to provide a reasonable basis for our findings and conclusions based on our audit objectives. We believe that the evidence obtained provides a reasonable basis for our findings and conclusions based on our audit objectives.

---

[8] For purposes of our report, we use the term recipient to include all entities (e.g., universities, industry) that received financial assistance for advanced fossil energy R&D projects regardless of the instrument used (e.g., by grant or cooperative agreement). We use university to include any institution of higher education, such as a public or non-profit private college, junior college, or university. We defined industry as any private sector organization that is not an institution of higher education, such as a consulting group or business. For more information on our methodology, see appendix I.

BACKGROUND

This section provides an overview of 1) DOE's administration of its advanced fossil energy R&D program, and 2) DOE's Loan Guarantee Program (LGP).

DOE's Administration of Its Advanced Fossil Energy R&D Projects

Within DOE, FE carries out DOE's program for fossil energy R&D, which includes federal research, development, and demonstration efforts on advanced power generation; power plant efficiency; water management; and carbon capture and storage (CCS) technologies. CCS is a process that involves capturing man-made CO_2 at its source and storing it permanently underground.[9] The program for fossil energy R&D also includes the development of technological solutions for the development of U.S. unconventional oil and gas domestic resources, such as from shale formations. FE also oversees the operations, infrastructure, and R&D at NETL, among other things. NETL officials told us that NETL has dual roles: it serves as project manager for advanced fossil energy R&D projects that receive federal assistance, and, as a DOE national laboratory, it also conducts applied research.[10] FE and NETL collaborate on the selection and administration of the awards for advanced fossil energy R&D projects, according to DOE officials.

DOE's efforts to administer its program for advanced fossil energy R&D take place across a spectrum of activities, including providing

---

[9] CCS is sometimes referred to as CCUS—carbon capture, utilization, and storage. In addition, CCS is also sometimes referred to as carbon capture and sequestration. CCS is also applicable to non-coal industrial processes such as corn ethanol plants and oil refineries.

[10] NETL is the only DOE national laboratory operated by the government rather than a contractor. NETL's in-house R&D work was outside of the scope of our review.

financial assistance for large demonstration projects.[11] In the 1980s and early 1990s, DOE's fossil energy R&D program primarily focused on reducing emissions of harmful pollutants from coal-fired power plants, particularly sulfur dioxide and nitrogen oxide. For example, DOE began its large demonstration projects of advanced coal technologies in the mid-1980s; this work focused on R&D to mitigate acid rain and to reduce the pollutants released from coal combustion. More recently, DOE has provided funding for advanced fossil energy R&D to reduce CO_2 emissions by developing beneficial uses for CO_2 from coal-fired power plants, and to improve methods for CCS, among other things. As we have previously reported, CCS is a key technology that shows potential for reducing CO_2 emissions from coal plants.[12] Specifically, CCS technologies separate and capture CO_2 from other gases produced when combusting or gasifying coal, compress it, then transport it to underground geologic formations such as saline aquifers—porous rock filled with brine—or oil and natural gas reservoirs, where the captured CO_2 is injected for long-term storage.

Globally, two power plants currently generate electricity from fossil fuel while capturing CO_2 in large quantities: the Boundary Dam plant in Canada and the Petra Nova plant in Texas. Both plants retrofitted CCS technology to existing plants. A third fossil-fueled, electricity-generating operation, the Kemper County Energy Facility in Mississippi, was scheduled to begin CCS operations in 2016, but cost overruns and delays in construction and operations led to the suspension of the plant's CCS component in June 2017. Each of these power plants using CCS systems may be described as a first-of-its-kind venture, using technologies developed at a pilot scale ramped up to commercial scale. It is not unusual for projects in the demonstration phase of the R&D process to experience

---

[11] Demonstration projects operate new or improved technologies to collect information on their performance and assess readiness for commercialization and deployment for widespread use.

[12] GAO, *Coal Power Plants: Opportunities Exist for DOE to Provide Better Information on the Maturity of Key Technologies to Reduce Carbon Dioxide Emissions*, GAO-10-675. (Washington, D.C.: June 16, 2010).

higher-than-anticipated costs, delays, and other challenges, according to a 2017 Congressional Research Service report.[13]

DOE generally uses announcements of opportunities for federal financial assistance to competitively solicit potential applicants of advanced fossil energy R&D projects. According to DOE officials, the department sets priorities for its advanced fossil energy R&D funding each year based in part on the amount appropriated for FE R&D and on FE's R&D plans, as well as any direction that Congress may have specified for certain types of technology R&D. DOE's advanced fossil energy R&D projects typically lasted for multiple years.

DOE sets milestones for technical progress for each year of a project to ensure that funding recipients accomplish a specific R&D objective or set of objectives, according to DOE officials. The recipient may submit some form of report on its progress on the R&D as well as accomplishments to DOE for review and approval to continue. DOE officials told us they review the progress of the recipient at each phase and the project continuation is subject to the recipient's technical progress, the recipient's compliance with all of the other terms—including any financial terms—of the agreement, and the availability of DOE's funds, based on congressional appropriations.

DOE's Loan Guarantee Program

The LGP was originally designed to address a fundamental impediment to innovative and advanced energy projects: securing enough affordable financing to survive the period between developing innovative technologies and commercializing them. As we have previously reported, these projects have risks, such as technology risk—the risk that the new technology will not perform as expected—and execution risk—the risk that

[13] Congressional Research Service, *Carbon Capture and Sequestration (CCS) in the United States*. (Washington D.C.: November 14, 2017).

the borrower or project will not perform as expected.[14] Because the risks that commercial lenders must assume to support new technologies can put the cost of private financing out of reach, companies may not be able to commercialize innovative technologies without the federal government's financial support.

Federal loan guarantee programs such as the LGP can help companies obtain financing because the federal government agrees to reimburse the lender for the guaranteed amount if a borrower defaults. Section 1703 of EPAct authorizes DOE to provide loan guarantees for projects that avoid, reduce, or sequester air pollutants or man-made emissions of greenhouse gases and employ new or significantly improved technologies as compared to commercial technologies in service in the United States at the time the guarantee is issued.[15] EPAct describes several categories of projects that are eligible for guarantees under the program, including, among others, renewable energy systems, efficient end-use energy technologies, advanced nuclear facilities, advanced fossil energy technology, and CCS technologies. DOE's Loan Programs Office, which administers the LGP, had issued three loan guarantees under Section 1703 supporting nuclear technologies as of August 2018, but none supporting advanced fossil energy or any other technologies.

DOE Provided $2.66 Billion in Funding for 794 Advanced Fossil Energy R&D Projects Started from Fiscal Years 2010 through 2017

DOE provided $2.66 billion in funding for 794 advanced fossil energy R&D projects started from fiscal years 2010 through 2017. These 794

---

[14] GAO, *DOE Loan Programs: Current Estimated Net Costs Include $2.2 Billion in Credit Subsidy, Plus Administrative Expenses*, GAO-15-438 (Washington, D.C.: April 27, 2015).

[15] Pub. L. No. 109-58, § 1703, 119 Stat. 594, 1120 (2005) (codified as amended at 42 U.S.C. § 16513).

projects included 9 later-stage large demonstration projects and 785 other advanced fossil energy R&D projects.[16] DOE provided $1.12 billion in funding to nine large projects aimed at demonstrating the commercial viability of CCS technologies. DOE provided $1.54 billion in funding to 785 other R&D projects for both coal and oil and gas technologies, mostly to universities and industry, located in 46 states and the District of Columbia.

DOE Provided $1.12 Billion in Funding for Nine Large Demonstration Projects Started from Fiscal Years 2010 through 2017

For nine large demonstration projects started from fiscal years 2010 through 2017, DOE provided $1.12 billion in funding. These projects received that funding from appropriations from the American Recovery and Reinvestment Act of 2009 (Recovery Act)[17] and supported efforts to reduce the financial and technical risks of commercial CCS, according to a 2017 report by the Congressional Research Service.[18] Six demonstration projects researched CCS technologies using coal, while three used other fuels, namely methane, ethanol, and petcoke.[19] Recipients were generally required to provide a certain percentage of the cost of each R&D project, called cost share.[20] Specifically, to receive funding, recipients of funding

[16] The 785 R&D projects include all the advanced fossil energy R&D projects except for the 9 large demonstration projects in our scope.

[17] The Recovery Act was enacted in 2009 to preserve and create jobs, promote economic recovery, and to provide investments needed to increase economic efficiency by spurring technological advances in science and health, among other things. Pub. L. No. 111-5, 123 Stat. 115 (2009).

[18] Congressional Research Service, *Carbon Capture and Sequestration: Research, Development, and Demonstration at the U.S. Department of Energy*. Washington D.C.: April 24, 2015.

[19] Petcoke—petroleum coke—is a waste byproduct of oil refining with an appearance and texture similar to coal.

[20] In general, once a project begins, the recipient is responsible for cost overruns, plus all the unallowable project costs (i.e., project costs in which the government is not permitted to share), according to DOE officials.

for the nine large demonstration projects agreed to pay at least $610 million in cost share for the demonstration projects. Three of those demonstration projects remained active at the end of fiscal year 2017. Four projects had their support withdrawn by DOE, and two were withdrawn by the recipient. These projects ended due to several factors such as a lack of technical progress, the closure of the Recovery Act appropriations account on September 30, 2015, and changing economic conditions—such as decreased natural gas prices which resulted in changes in the relative prices of coal and natural gas.

The nine large demonstration projects represented over 40 percent of the $2.66 billion in advanced fossil energy R&D funding for the 794 projects (see Figure 1).

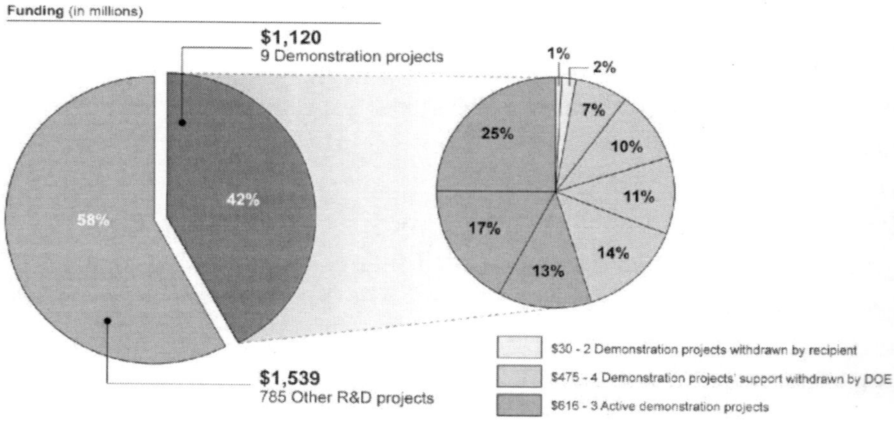

Source: GAO analysis of Department of Energy (DOE) data. | GAO-18-619.

Note: Of the 794 research and development projects, 9 later-stage large demonstration projects supported carbon capture and storage technologies to assess their readiness for commercial viability. Advanced fossil energy projects include research on innovative coal, natural gas, and oil technologies. For purposes of our report, we use the term funding to mean obligations. Funding amounts may not sum exactly due to rounding.

Figure 1. Distribution of Department of Energy (DOE) Funding for Advanced Fossil Energy Research and Development Projects that Started from Fiscal Years 2010 through 2017.

Three Large Demonstration Projects Remained Active as of the End of Fiscal Year 2017

Of the $1.12 billion in funding for the advanced fossil energy demonstration projects, DOE provided $616 million in funding for three large demonstration projects that started in fiscal year 2010 and that remained active as of the end of fiscal year 2017.

- Petra Nova Parish Holdings of Texas has a demonstration project underway that has retrofitted an existing coal-fired power plant in Texas with post-combustion carbon capture technology, according to DOE documentation.[21] The objective of this project is to demonstrate the ability to capture 90 percent of the CO_2 emitted from a flue gas stream—gas created from burning coal. DOE provided $190 million in funding for the project from fiscal years 2010 through 2017.[22] According to DOE documentation, DOE's involvement with the project is scheduled to conclude in December 2019. The Petra Nova project captured and stored its first 1 million metric tons of CO_2 in November 2017, according to DOE officials.
- Archer Daniel Midlands of Illinois had a demonstration project underway to capture CO_2 produced as a by-product of the production of fuel-grade ethanol for its demonstration project in Illinois, according to DOE documentation. The objective of this project is to capture approximately 1 million tons of CO_2 per year using dehydration and compression processes and sequester it in the Mt. Simon Sandstone formation (a saline reservoir) in

[21] Post-combustion carbon capture is a process in which the CO2 is captured from flue gases (gases that are emitted as exhaust) after the fossil fuel has been burned. It is used mainly for conventional coal-fired power generation, but the term can also apply to combustion turbines and other combustion systems fired by natural gas.

[22] While our scope included advanced fossil energy R&D projects that started in fiscal years 2010 through 2017, DOE officials told us that in a few instances they provided some funding at the time of the award, in fiscal year 2009, prior to the start of the project.

Illinois.[23] DOE provided $141 million in funding for the project from fiscal years 2010 through 2017. DOE's involvement with the project is scheduled to conclude in September 2019, according to DOE documentation. During calendar year 2017, the project captured and stored over 500,000 metric tons of CO_2, according to DOE officials.

- Air Products and Chemicals of Pennsylvania designed, constructed, and began operating a system to capture the CO_2 emitted from two large steam-methane reformers, which produce hydrogen from methane, for its demonstration project in Texas.[24] The captured gas is compressed and sent via pipeline to oil fields in eastern Texas to be used for enhanced oil recovery and thereby sequestered, according to DOE documentation.[25] DOE provided $284 million in funding for the project from fiscal years 2010 through 2017. DOE's involvement under this demonstration project's award concluded the last day of fiscal year 2017.[26]

DOE Withdrew Support for Four Large Demonstration Projects

DOE provided assistance for four large demonstration projects until the agency withdrew support for them. Of the total $1.12 billion in funding for the nine advanced fossil energy demonstration projects, DOE provided $475 million in funding for these four projects. Demonstration projects are

---

[23] A saline reservoir is a subsurface saline aquifer (water-bearing stratum of permeable sand, rock, or gravel).

[24] Most hydrogen produced today in the United States is made via steam-methane reforming, a mature production process in which high-temperature steam (700°C– 1,000°C) is used to produce hydrogen from a methane source, such as natural gas.

[25] Enhanced oil recovery is the process in which heat, chemicals, or gases such as CO2 are injected into depleted oil and gas formations to recover those resources that would otherwise be difficult to reach. Gas injection accounts for nearly 60 percent of enhanced oil recovery production in the United States.

[26] As of May 1, 2018, the project was still in operation and had captured and stored 4.5 million metric tons of CO_2, according to DOE officials.

inherently risky, and DOE withdrew support for these projects for a number of reasons, including:

- DOE provided financial assistance for two large demonstration projects through FutureGen 2.0 that started at the beginning of fiscal year 2011.[27] The first project was expected to use oxy-combustion technology to create a near-zero emissions coal-fired power plant by capturing and compressing at least 90 percent of the plant's CO_2 and eliminating almost all the sulfur oxides, nitrogen oxides, particulate, and mercury pollutants from plant emissions, according to DOE documentation.[28] The second project concurrently planned to construct and operate a CO_2 pipeline from the plant site to a storage site located about 30 miles to the east where approximately 1 million metric tons per year of compressed and purified CO_2 were to be injected into the Mt. Simon Sandstone formation (a saline reservoir) in Illinois. DOE provided just over $200 million in total funding for both projects. According to a DOE fact sheet on FutureGen 2.0, DOE initially estimated the cost expected to be borne by the federal government for the completion of both projects to be over $1 billion. DOE directed the suspension of FutureGen 2.0 project development activities in February 2015 because DOE concluded that there was insufficient time to

---

[27] In 2003, DOE announced its FutureGen program as a $1 billion venture, partnering with the electric power industry to design, build, and operate the world's first coal-fired, zero-emissions power plant. In 2005, the FutureGen Industrial Alliance, a nonprofit consortium of some of the largest coal producers and electric power companies in the world, formed to join DOE in this effort. In January 2008, DOE announced that it would not continue its cooperative agreement with the Alliance and that it was going to take a different approach to FutureGen. The original FutureGen was a DOE research and development project, but the restructured FutureGen was a DOE commercial demonstration project. For more information, see GAO, *Clean Coal: DOE's Decision to Restructure FutureGen Should Be Based on a Comprehensive Analysis of Costs, Benefits, and Risks*, GAO-09-248 (Washington, D.C.: Feb. 13, 2009).

[28] Oxycombustion is a process in which coal is burned in oxygen instead of in air, with the resulting exhaust containing only CO2 and water vapor. Because it yields an almost 100-percent CO2 stream that is readily transportable, the process has strong potential but is extremely energy-intensive, according to DOE documentation.

complete the projects before the closure of the Recovery Act appropriations account on September 30, 2015.

- DOE provided assistance for one large demonstration project started in fiscal year 2010 by Hydrogen Energy California. This project was to design, build, and operate a commercial scale, advanced integrated gasification combined cycle power plant and fertilizer production facility with CCS in California.[29] The project was expected to achieve at least 90 percent carbon capture efficiency while storing approximately 2.6 million tons of CO_2 per year in an enhanced oil recovery application. DOE provided a total of $154 million in funding for the project. According to a DOE fact sheet on this project, DOE initially estimated the cost expected to be borne by the federal government for the completion of this project to be $408 million. However, according to DOE officials, the recipient faced a number of challenges, such as obtaining permits, planning a railroad extension, and securing additional financing sources for construction. The recipient repeatedly missed milestones under the award that delayed the project's ability to meet DOE's objectives, according to DOE officials. As a result of these delays, and the closure of the Recovery Act appropriations account on September 30, 2015, DOE withdrew funding for this project after the recipient's period of performance ended on January 20, 2015.
- DOE provided assistance for one large demonstration project started in fiscal year 2010 by Summit Texas Clean Energy. This project was to integrate carbon capture technology with a new integrated gasification combined cycle power facility in Penwell, Texas. Captured CO^2 was to be sold for various enhanced oil

[29] In integrated gasification combined cycle (IGCC) power plants, coal is gasified to produce synthesis gas, consisting primarily of hydrogen, carbon monoxide, and CO2. Through IGCC, electricity is generated more efficiently than through conventional pulverized coal-fired technology—the process most widely in use—because IGCC uses less coal to generate the same amount of electricity. In addition, oxygen-fired IGCC plants produce CO2 as a concentrated gas stream at high pressure that may be captured and stored more easily and cheaply than CO2 from a typical pulverized coal-fired power plant, which emits CO2 that must be separated from other gases before storing.

recovery operations. The project was originally slated to have a groundbreaking in mid- to late-summer 2015. Construction was expected to be completed in mid-2018 with a ramp up to full commercial operations to begin soon afterward. DOE provided $120 million in funding for the project. According to a DOE fact sheet about this project, DOE initially estimated the cost expected to be borne by the federal government for the completion of this project to be $450 million. In 2017, DOE withdrew funding for this demonstration project because the recipient had failed to achieve the technical objectives specified under the DOE award and had not secured financing for the project, according to DOE officials. The DOE Office of Inspector General issued a special report on this project in 2016 to discuss its concern about the viability of the project.[30] In 2018, after DOE had withdrawn its support for the project, the Inspector General issued a full audit report, stating that DOE had not always effectively and efficiently managed financial aspects of this project.[31] According to DOE officials, DOE management accepted the Inspector General's audit recommendations and initiated steps to address them.

Two Recipients Withdrew Their Projects

DOE provided assistance for two large demonstration projects from which the recipients subsequently withdrew. Of the total $1.12 billion in funding for the advanced fossil energy demonstration projects, DOE provided a total of $30 million in funding for these two demonstration projects.

---

[30] Department of Energy, Office of Inspector General, *The Department of Energy's Continued Support of the Texas Clean Energy Project Under the Clean Coal Power Initiative*, OIG-SR-16-02 (Washington, D.C.: Apr. 26, 2016).

[31] Department of Energy, Office of Inspector General, *The Office of Fossil Energy's Oversight of the Texas Clean Energy Project Under the Clean Coal Power Initiative*, DOEOIG-18-17 (Washington, D.C.: Feb. 8, 2018).

- American Electric Power of West Virginia, starting in fiscal year 2010, was to lead the design, construction, and operation of a CCS system using the chilled ammonia process, which was expected to capture at least 90 percent of the CO_2 from the Mountaineer coal-fired power plant located near New Haven, West Virginia.[32] Captured CO_2 was to have been permanently sequestered in a subsurface geologic formation. DOE provided $17 million in funding for the project. According to a DOE fact sheet about this project, DOE initially estimated the cost expected to be borne by the federal government for the completion of this project to be $334 million. The project ended on September 30, 2011 with the completion of all phase I milestones and deliverables. The recipient withdrew after completion of phase I and terminated this demonstration project before advancing to the design, construction, and operations stages. This decision to withdraw was the result of challenges that the recipient faced in its ability to fund its portion of the cost share for the commercial scale project, among other things, according to DOE documentation.
- Leucadia Energy of New York, starting in fiscal year 2010, was to demonstrate the capture of CO_2 from an industrial facility for use in an independent, enhanced oil recovery application. The industrial source of CO_2 was to be a petcoke-to-chemicals (i.e., methanol and other by-products) gasification plant being developed by Lake Charles Clean Energy in Lake Charles, Louisiana.[33] DOE provided $13 million in funding for the project. According to a DOE fact sheet about this project, DOE initially estimated the cost expected to be borne by the federal government

---

[32] In the chilled ammonia process, an ammonia based solution reacts with CO2 of cooled flue gas. Raising the temperature reverses this reaction and pressurized CO2 is released. The solution is then recycled.

[33] Methanol is one of the world's most widely-used industrial chemicals in applications including paints and plastics, furniture and carpeting, automotive parts, windshield washer fluid, and fuel blending.

for the completion of this project to be $261 million. On September 24, 2014, due to its business assessment of the core methanol plant, Leucadia Energy informed DOE that it would withdraw and not pursue the main petcoke-to-methanol project, and as such could not continue with the proposed demonstration project. DOE officials told us this decision was unrelated to the additional deployment of the CCS technologies that were supported by the DOE demonstration project, which ended at that point.[34]

DOE Provided $1.54 Billion in Funding for 785 Other R&D Projects Started from Fiscal Years 2010 through 2017 to Support Advanced Fossil Energy

DOE provided $1.54 billion in funding for 785 other advanced fossil energy R&D projects started from fiscal year 2010 through 2017. For these 785 R&D projects, DOE provided:

- on average, $2.0 million per project;
- a median of $0.8 million per project;
- less than $5 million to 91.8 percent (721) of the 785 projects; and
- less than $1 million to 58.1 percent (456) of the projects.

For projects started from fiscal years 2010 through 2017, total funding for projects by fiscal year started ranged from less than $100 million to more than $300 million (see Figure 2).

[34] On December 21, 2016, DOE announced a conditional commitment for a loan guarantee of up to $2 billion via its Loan Programs Office's LGP to the Lake Charles Methanol Project, which would leverage upon the work and experience gained from the earlier Leucadia Energy project. We discuss the Lake Charles Methanol Project later in this report.

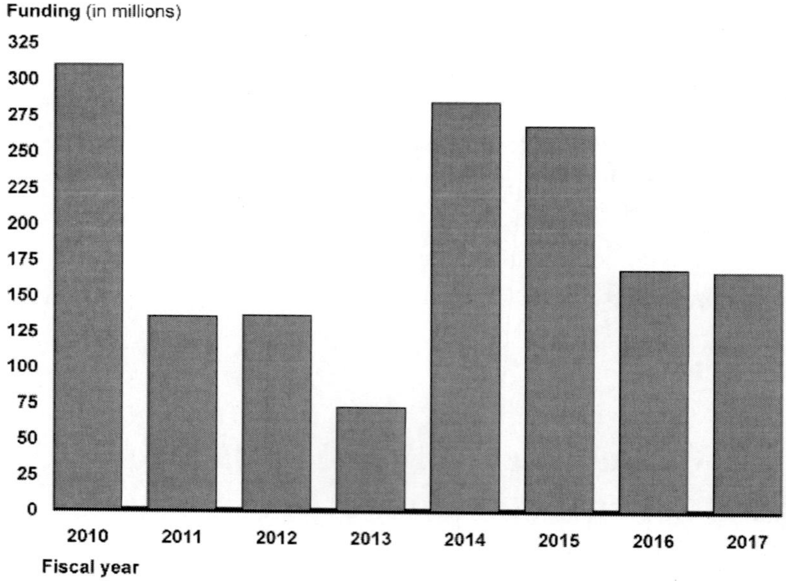

Source: GAO analysis of DOE data. | GAO-18-619.

Note: Advanced fossil energy projects include research on innovative coal, natural gas, and oil technologies. For purposes of our report, we used the term funding to mean obligations, and we defined the 785 R&D projects to be inclusive of the advanced fossil energy R&D projects except for the 9 large demonstration projects in our scope. This bar chart includes 785 research and development projects—that is, all of the projects in our scope that were not the large demonstration projects. The data for each fiscal year represent total obligations for the projects that started in that fiscal year, rather than total obligations by fiscal year.

Figure 2. Department of Energy's (DOE) Funding for Advanced Fossil Energy Research and Development (R&D) Projects Other than Large Demonstration Projects Started from Fiscal Years 2010 through 2017, by Fiscal Year Started.

As noted earlier, recipients of DOE's R&D funding were generally required to provide cost share to support the cost of each R&D project. For 661 of the 785 projects, the initially agreed-upon dollar amount to be covered by recipients was $617 million in cost-share. Recipients did not provide a cost-share for the remaining 124 of the 785 projects, which were predominantly grants without cost share requirements, according to DOE officials.

According to DOE data, DOE provided the largest amount of funding for projects started in 2010 because DOE received a supplemental appropriation for fossil energy R&D through the Recovery Act. DOE provided funding for 72 of the coal technologies research projects—totaling $237 million—using appropriations from the Recovery Act, according to DOE data.

Of the 785 R&D projects for which DOE provided funding, most advanced fossil energy projects researched coal technologies rather than oil and gas, and recipients of the funding were generally universities and industry groups that were distributed across the country.

Most of DOE's 785 Advanced Fossil Energy Projects Researched Coal Technologies

Of the 785 projects, 698 (about 89 percent) involved coal technologies, receiving $1.40 billion (about 91 percent) of the $1.54 billion in funding DOE provided for the projects. The remaining projects and funding supported R&D for oil and gas technologies, according to DOE's categorization of the projects by fuel type (see Table 1).

Table 1. Department of Energy's (DOE) Funding for Advanced Fossil Energy Research and Development (R&D) Projects Started from Fiscal Years 2010 through 2017 by Fuel Type

Fuel type[a]	Number of projects	Funding amounts (in dollars)
Coal	698	1,396,712,035
Oil and Gas	87	141,888,174
Total	785	1,538,600,209

Source: GAO analysis of DOE data. | GAO-18-619.

Note: Advanced fossil energy projects include research on innovative coal, natural gas, and oil technologies. For purposes of our report, we used the term funding to mean obligations, and we defined the 785 R&D projects to be inclusive of the advanced fossil energy R&D projects except for the 9 large demonstration projects in our scope.

[a]DOE identified the fuel types for each project. This table represents funding for R&D projects provided through DOE's National Energy Technology Laboratory.

Within each fuel type, projects researched various technology types, such as R&D on coal gasification systems and the mitigation of methane emissions from natural gas infrastructure.[35] The funding for the 785 R&D projects ranged from $5,000 for a research conference (oil and gas) to $125 million for a research facility focused on next-generation CCS technologies (coal).

The funding for the 698 R&D projects that are coal-related ranged from $28,000 for detection of low concentrations of heavy metals in power plant water supplies to the active $125 million project for a research facility focused on next-generation CCS technologies. In a 2017 report, DOE identified the following six categories as coal-related research areas for its R&D projects:[36]

- **Carbon Capture:** The development of research to enable the cost-effective implementation of carbon capture technologies throughout the power generation sector, including at new and existing coal-based power plants. For example, DOE provided FuelCell Energy Inc., of Connecticut, about $4 million in funding for a 3-and-a-half-year active project for a pilot test of a novel system for CO_2 capture and power generation.
- **Carbon Storage:** The development of processes of storing the CO_2 that has been separated and captured from coal-powered processes in deep underground geological formations. For example, DOE provided the University of Texas at Austin $1 million in funding for a 1-and-a-half-year project to study the feasibility of integrated CCS in the northwest Gulf of Mexico.
- **Advanced Energy Systems:** The development of energy technologies that can minimize the environmental impact of coal use and optimize energy resources. DOE has provided financial assistance for R&D in this category for the process of converting

---

[35] Coal gasification is the process in which carbon-containing material is converted into synthesis gas.

[36] Department of Energy, *National Energy Technology Laboratory (NETL) Coal Research and Development: Overview* (Aug. 15, 2017).

carbon-containing material (i.e., coal) into synthesis gas, and the development of advanced turbines that will serve to improve the efficiency and reduce the cost of coal-fueled power plants with carbon capture, among other areas. For example, DOE provided Ohio State University over $600,000 in funding for a 3-year active project on revolutionizing turbine cooling.

- **Supercritical CO_2 Technology:** The development of directly-heated power cycles using more advanced coal conversion systems based on supercritical CO_2 cycles.[37] This technology seeks to use CO_2 as the working fluid in the turbomachinery compared to existing steam-based power cycles in coal plants. For example, DOE provided the Institute of Gas Technology Inc., of Illinois, $35 million in funding for a 6-year active project for a supercritical CO_2 pilot plant test facility planned to be completed at the end of fiscal year 2022.
- **Crosscutting Technologies:** The development of tools, support of research, and sponsorship of science and engineering education designed to contribute to, and encourage, greater collaboration among disciplines and across each of the coal-related research areas. DOE transitions successful crosscutting technologies to its other coal R&D programs. DOE has provided financial assistance for R&D in this category for the development of functional materials for advanced fossil energy power production technologies and the treatment of water extracted by injection of CO_2 into deep saline aquifers, among other areas. For example, DOE provided the Research Triangle Institute in North Carolina $750,000 in funding for a 2-year project on a low-cost, low-energy treatment process using a non-aqueous solvent extraction to recover water from deep aquifer brines.

---

[37] Supercritical power cycles are operated above the critical point of CO2 so that it does not change phases, but rather undergoes drastic density changes over small ranges of temperature and pressure. Such cycles have shown the potential for increased heat-to-electricity conversion efficiencies, high power density, and simplicity of operation compared to existing steam-based power cycles.

- **Rare Earth Elements:** The development of research and methods for rare earth elements' separation and recovery from coal and coal by-products, among other things. For example, DOE provided the University of Kentucky $400,000 in funding for a 1- and-a-half-year project to identify and characterize rare earth elements from coal and coal by-products.

The other 87 of the 785 projects (about 11 percent) researched oil and gas technologies, such as hydraulic fracturing (fracking) and the production of hydrocarbons—primarily natural gas—from shale formations. The funding for the oil and gas R&D projects ranged from $5,000 for a research conference to $29 million for the University of Texas at Austin's active project on the deep-water characterization and scientific assessment of gas hydrates. Specifically, DOE identified the following four categories as oil and gas-related research areas:

- **Gas Hydrates:** The development of technologies to find, characterize, and recover methane from gas hydrates through field testing, numerical simulation, and laboratory experimentation, among other things. For example, DOE provided the University of California-San Diego $350,000 in funding for a 3-year active project to characterize the baselines and changes in gas hydrate systems.
- **Natural Gas Infrastructure:** The monitoring of the U.S. natural gas pipeline network, which includes more than 300,000 miles of interstate and intrastate transmission pipelines. For example, DOE provided the University of Pittsburgh $1.2 million in funding for a 3-year active project on multi-functional fiber sensors for pipeline monitoring and methane detections.
- **Onshore Unconventional Resources:** The production of hydrocarbons–primarily natural gas–from shale formations. For example, DOE provided the Ground Water Protection Council, of Oklahoma, $13 million for an 8-year project for data management

and regulatory approaches related to hydraulic fracturing and geologic sequestration of CO_2.

Offshore Oil and Gas

Although this research area is not currently active, according to DOE documentation, the multiyear oil and gas R&D projects that received funding prior to 2014 continued through September 2016.[38] The R&D projects in this area included research on geologic uncertainty prediction of oil and gas, and improvement of subsea systems reliability through automation and advanced technology.

Recipients Were Generally Universities and Industry Groups That Were Distributed Across the Country

The recipients of the funding for the 785 advanced fossil energy R&D projects were mostly universities and industry groups that were located in 47 states and the District of Columbia. Of these recipients, approximately 51 percent were universities; 43 percent were industry groups; and 5 percent were other entities, including other federal agencies, such as the U.S. Geological Survey (see Table 2).

While university recipients received funding for a majority of projects, industry recipients received a majority of the funding (see Table 3).

Recipients were located in 47 states and the District of Columbia.[39] The three states with the highest number of projects with recipients located in their states were Texas (100), California (61), and Ohio (53). The three states where recipients received the most funding were Texas (about $169

---

[38] Section 301 of the Bipartisan Budget Act of 2013 repealed provisions of the Energy Policy Act of 2005 that had authorized an R&D program relating to ultra-deepwater and unconventional natural gas and other petroleum resource exploration and production. Pub. L. No. 113-67, § 301, 127 Stat. 1165, 1181 (2013) (repealing Pub. L. No. 109-58, Title IX, Subtitle J, 119 Stat. 594, 916 (2005)). The 2013 law also rescinded any unobligated funds appropriated for carrying out that program.

[39] This information refers to the location of the entity that received funding. Projects did not necessarily take place in the states where recipients were located. Additionally, according to DOE officials, four recipients were located outside the United States.

million), Alabama (about $161 million), and California (about $152 million) (see Figure 3).

Table 2. Recipient Types for the Department of Energy's (DOE) Advanced Fossil Energy Research and Development (R&D) Projects Started from Fiscal Years 2010 through 2017

Recipient type	Number of recipients	Percent of all recipients
University[a]	165	51
Industry[b]	140	43
Other[c]	17	5
Total	322	100

Source: GAO analysis of DOE data. | GAO-18-619.

Note: Advanced fossil energy projects include research on innovative coal, natural gas, and oil technologies. For purposes of our report, we defined the 785 R&D projects to be inclusive of the advanced fossil energy R&D projects except for the 9 large demonstration projects in our scope. Percentages may not sum to 100 due to rounding.

[a]We defined university as any institution of higher education, such as a public or non-profit private college, junior college, or university.

[b]We defined industry as any private sector organization that is not an institution of higher education, such as a consulting group or business.

[c]Other includes any entity not associated with a university or industry, including federal agencies.

Table 3. Department of Energy's (DOE) Advanced Fossil Energy Research and Development (R&D) Projects and Funding by Recipient Type for Projects Started during Fiscal Years 2010 through 2017

Recipient type	Number of projects	Percent of all projects	Funding amounts (in dollars)	Percent of all funding
University[a]	466	59	566,138,094	37
Industry[b]	289	37	922,989,703	60
Other[c]	30	4	49,472,412	3
Total	785	100	1,538,600,209	100

Source: GAO analysis of DOE data. | GAO-18-619.

Note: Advanced fossil energy projects include research on innovative coal, natural gas, and oil technologies. For purposes of our report, we used the term funding to mean obligations, and we defined the 785 R&D projects to be inclusive of the advanced fossil energy R&D projects except for the 9 large demonstration projects in our scope.

[a]We defined university as any institution of higher education, such as a public or non-profit private college, junior college, or university.

[b]We defined industry as any private sector organization that is not an institution of higher education, such as a consulting group or business.

[c]Other includes any entity not associated with a university or industry, including federal agencies.

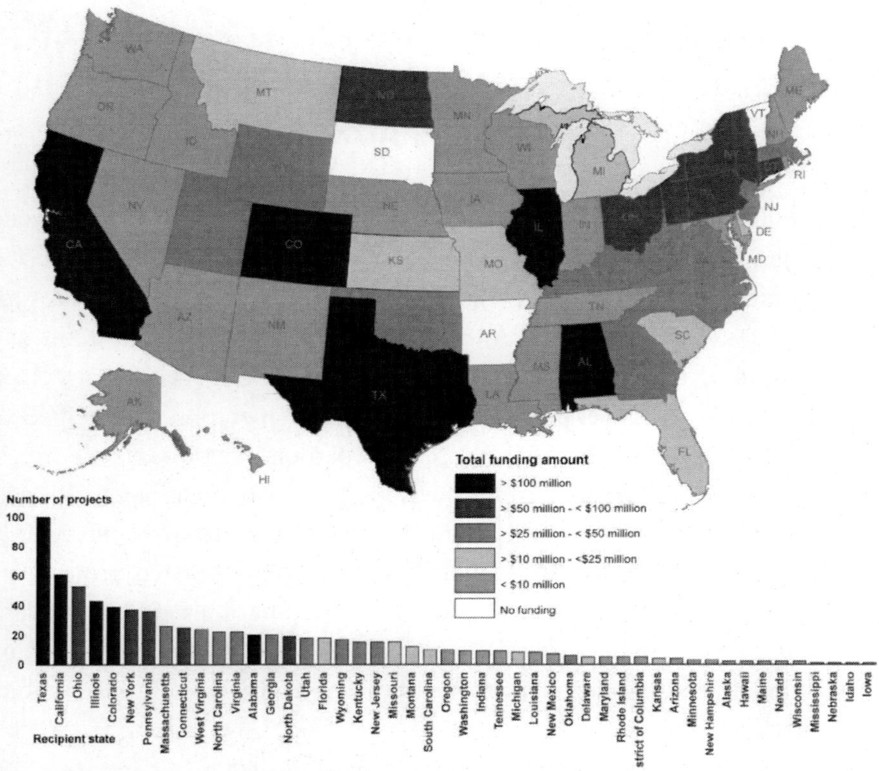

Source: GAO analysis of DOE data, Map Resources (map). | GAO-18-619.

Note: This information refers to the location of the entity that received funding. Projects did not necessarily take place in the state where their recipient was located. Additionally, according to DOE officials, four recipients were located outside the United States. Advanced fossil energy projects include research on innovative coal, natural gas, and oil technologies. For purposes of our report, we used the term funding to mean obligations, and we defined the 785 R&D projects to be inclusive of the advanced fossil energy R&D projects except for the 9 large demonstration projects in our scope.

Figure 3. Department of Energy's (DOE) Advanced Fossil Energy Funding for Research and Development (R&D) Projects Started from Fiscal Years 2010 through 2017 by Recipient Location.

DOE MADE NO LOAN GUARANTEES FOR ADVANCED FOSSIL ENERGY FROM FISCAL YEAR 2006 THROUGH AUGUST 2018

Although DOE issued three solicitations for applications for advanced fossil energy loan guarantees—most recently in fiscal year 2014, for up to $8 billion in loan guarantees—DOE had not guaranteed any loans for advanced fossil energy as of August 2018.[40] Specifically, the 2006 and 2008 advanced fossil energy solicitations were for projects that involved coal-based power generation and that would incorporate CCS, coal gasification, or other beneficial uses of carbon, among other things. However, neither solicitation resulted in any loan guarantees, in part because during this timeframe of the late 2000s, natural gas prices fell, causing a shift in the market, which led to such coal-related projects no longer being economically competitive, according to DOE officials.[41] According to the fiscal year 2014 solicitation, applicants could use any fossil fuel—including coal, oil, or natural gas—that would reduce, avoid, or sequester greenhouse gases.

In response to the 2014 advanced fossil energy solicitation, DOE officials told us that DOE had received 19 applications total. According to DOE officials:

- Five fossil energy applicants were actively moving through the process of review as of August 2018. For example, in January 2018, one applicant issued a press release stating that it was pursuing a $1.9 billion loan guarantee to support the development of infrastructure for a proposed underground storage facility for natural gas liquids and intermediates.

---

[40] DOE also issued advanced fossil loan guarantee solicitations in 2006, for up to $2 billion, and in 2008, for up to $8 billion. Both prior solicitations have closed. The 2014 solicitation will close in fiscal year 2020.

[41] According to DOE documentation, DOE received a total of 26 applications for the 2006 advanced fossil energy solicitation, and 8 applications for the 2008 solicitation.

- Nine fossil energy applicants had been idle or not following up with the Loan Programs Office.
- Three applicants did not meet certain eligibility requirements.[42]
- Two companies withdrew their applications—one in 2014, and one in 2018.

Of the five advanced fossil energy applicants actively in the process of DOE review, DOE offered a conditional commitment to guarantee up to $2 billion in loans to one applicant—Lake Charles Methanol—in December 2016. As we have previously reported, a conditional commitment is one where DOE commits to issue a loan guarantee if the applicant satisfies specific requirements.[43] According to information on the DOE website, the Lake Charles Methanol plant in Louisiana would produce methanol from the gasification of petcoke, and capture and transport the CO_2 to Texas for enhanced oil recovery. According to DOE documentation, the Lake Charles project planned to leverage the work and experience gained from the earlier DOE demonstration project by Leucadia Energy.[44]

AGENCY COMMENTS

We provided a draft of this report to DOE for review and comment. DOE provided technical comments, which we incorporated as appropriate.

As agreed with your office, unless you publicly announce the contents of this report earlier, we plan no further distribution until 30 days from the

---

[42] For example, in order to receive loan guarantees under Section 1703 of EPAct, including for advanced fossil energy, projects must meet certain greenhouse gas requirements. Specifically, to be eligible for loan guarantees, projects must avoid, reduce, or sequester air pollutants or man-made emission of greenhouse gases; and must employ new or significantly improved technologies as compared to commercial technologies in service in the United States at the time of loan guarantee issuance. 42 U.S.C. § 16513(a).

[43] The Secretary of Energy has the discretion to cancel a conditional commitment at any time for any reason prior to the issuance of a loan guarantee. See GAO, *DOE Loan Guarantees: Further Actions Are Needed to Improve Tracking and Review of Applications*, GAO-12-157 (Washington, D.C.: Mar. 12, 2012).

[44] Leucadia received funding from DOE as a large demonstration project, as noted previously.

report date. At that time, we will send copies to the appropriate congressional committees, the Secretary of Energy, and other interested parties. In addition, the report will be available at no charge on the GAO website at http://www.gao.gov.

If you or your staff members have any questions about this report, please contact me at (202) 512-3841 or ruscof@gao.gov. Contact points for our Offices of Congressional Relations and Public Affairs may be found on the last page of this report. GAO staff who made major contributions to this report are listed in appendix II.

Sincerely yours,

Frank Rusco
Director, Natural Resources and Environment

APPENDIX I: OBJECTIVES, SCOPE, AND METHODOLOGY

In this report, we describe 1) the Department of Energy's (DOE) funding of advanced fossil energy research and development (R&D) projects started from fiscal years 2010 through 2017 and the types of projects and recipients that received funding, and 2) DOE's loan guarantees, if any, for advanced fossil energy projects from fiscal year 2006 through August 2018.[45] You asked us to review DOE's funding for advanced fossil energy projects.

---

[45] For the purposes of this report, we use the term funding to mean obligations. An obligation is a definite commitment that creates a legal liability of the government for the payment of goods and services ordered or received, or a legal duty on the part of the United States that could mature into a legal liability by virtue of actions on the part of the other party beyond the control of the United States. Payment may be made immediately or in the future. An agency incurs an obligation, for example, when it places an order, signs a contract, awards a grant, purchases a service, or takes other actions that require the government to make payments to the public or from one government account to another. GAO, *A Glossary of Terms Used in the Federal Budget Process*, GAO-05-734SP (Washington, D.C.: Sept. 2005).

To address the first objective, we reviewed relevant laws, regulations, and DOE guidance. We analyzed DOE advanced fossil energy R&D project data for fiscal years 2010 through 2017. We focused our review on advanced fossil energy R&D projects that received funding through the Office of Fossil Energy's (FE) National Energy Technology Laboratory (NETL) because the 794 projects represent all of the advanced fossil energy R&D projects in our scope started from fiscal years 2010 through 2017. We used fiscal year 2010 as the start date because DOE officials told us that DOE's current data management system came into use for the R&D projects that started in fiscal year 2010.

We used fiscal year 2017 as the end date because that was the most recent complete year for which data were available. DOE provided us with a spreadsheet that included key project information—such as the name of the recipient of the R&D funding and the project start date—as well as obligations data for each project started for the period of our review (fiscal years 2010 through 2017), by the fiscal year during which the project was started, by summing the obligations for the project from each year. We reported on DOE's funding for these R&D projects; DOE generally provided financial assistance for these projects through grants or cooperative agreements.[46] Because advanced fossil energy R&D financial assistance for the Small Business Innovation Research and Small Business

[46] Federal financial assistance means assistance that non-federal entities receive or administer in the form of grants, property, cooperative agreements, food commodities, direct appropriations, or other assistance, and can also include loans, loan guarantees, interest subsidies, and insurance, depending on the context, but does not include amounts received as reimbursement for services rendered to individuals in accordance with OMB-issued guidance. 2 C.F.R. § 200.40. See also 31 U.S.C. § 7501(5). A grant agreement is generally defined as a legal instrument of financial assistance between a federal awarding agency and a non-federal entity that is used to enter into a relationship the principal purpose of which is to transfer anything of value from the federal awarding agency to the non-federal entity to carry out a public purpose authorized by law, and not to acquire property or services for the federal awarding agency's direct benefit or use. 2 C.F.R. § 200.51. A cooperative agreement is distinguished from a grant in that it provides for substantial involvement between the federal awarding agency and the non-federal entity in carrying out the activity contemplated by the federal award. 2 C.F.R. § 200.24. For purposes of our report, we use the term awards to refer to both grants and cooperative agreements.

Technology Transfer programs, as well as for the DOE national laboratories, are not obligated through NETL, we excluded these projects from our scope.[47] In addition, NETL's in-house R&D work was outside of the scope of our review. To assess the reliability of the funding data, as well as the specific project information for the 794 R&D projects, we interviewed data specialists at DOE Headquarters, FE, and NETL and reviewed DOE internal guidance for the maintenance of agency data. We found the data to be sufficiently reliable for our purposes. We also reviewed DOE websites and documentation, including fact sheets, and interviewed officials from FE and NETL.

To characterize the kinds of groups that received advanced fossil energy R&D funding, we developed the following definitions for coding each recipient:

- University: any institution of higher education, such as a public or non-profit private college, junior college, or university.
- Industry: any private sector organization that is not an institution of higher education, such as a consulting group or business. Industry includes any for-profit organizations involving a corporation, partnership, sole proprietorship or other organization which is organized primarily for profit. Industry includes some organizations that were founded as non-profit corporations but call themselves "companies" and/or describe "serving clients."
- Other: any entity not associated with a university or industry. Other includes groups such as other federal government agencies, as well as non-profit corporations and other entities which we could not identify conclusively as either industry or universities.

We used these three categories, and their definitions, to guide us in the coding process. After developing these definitions, three analysts

[47] We also did not include tax expenditures administered by the Department of the Treasury, such as the tax credits for industrial carbon dioxide capture and sequestration and investment in clean coal facilities.

independently coded each recipient as a university, industry, or other. Our method was to examine the identifying information on each recipient's website and decide which category best described the entity. We also had an independent analyst check the coding category that we had assigned to each recipient and verify that we had made a reasonable coding decision.

To describe the status of DOE's advanced fossil energy loan guarantees, we reviewed relevant laws, regulations, and guidance, as well as past GAO reports describing DOE's administration of the loan program. We also reviewed summary information that DOE provided on applications for loan guarantees for advanced fossil energy projects. We analyzed information that DOE provided on applications for loan guarantees for advanced fossil energy projects under the Loan Guarantee Program (LGP) and other related information for fiscal year 2006 through August 2018. We used fiscal year 2006 as the start date because it was the first year that DOE issued an advanced fossil energy project solicitation—an announcement of opportunities for loan guarantees for advanced fossil energy projects—and we used August 2018 as the end date in order to provide the most up-to-date information as possible. We also reviewed the advanced fossil energy project solicitations DOE issued during this timeframe. To assess the reliability of the summary information, we interviewed LGP staff who maintain the information for the advanced fossil energy applications, and reviewed DOE documentation. We found the data to be sufficiently reliable for our purposes. In addition, we interviewed officials from the Loan Programs Office who work on the LGP.

We conducted this performance audit from March 2017 to September 2018 in accordance with generally accepted government auditing standards. Those standards require that we plan and perform the audit to obtain sufficient, appropriate evidence to provide a reasonable basis for our findings and conclusions based on our audit objectives. We believe that the evidence obtained provides a reasonable basis for our findings and conclusions based on our audit objectives.

GAO Contact and Staff Acknowledgments

GAO Contact

Frank Rusco, (202) 512-3841 or ruscof@gao.gov

Staff Acknowledgments

In addition to the contact named above, Karla Springer (Assistant Director), Rebecca Makar (Analyst-in-Charge), TC Corless, Cindy Gilbert, Carol Henn, Kirk Menard, Patricia Moye, Sheryl Stein, and Sara Sullivan made key contributions to this report. Also contributing to this report were Carolyn Blocker, Marcia Carlsen, Nirmal Chaudhary, Jaci Evans, Ryan Gottschall, Keesha Luebke, and John Yee.

In: Advanced Fossil Energy …
Editor: Edward Carr
ISBN: 978-1-53615-071-1

Chapter 2

SCIENCE AND TECHNOLOGY: CONSIDERATIONS FOR MAINTAINING U.S. COMPETITIVENESS IN QUANTUM COMPUTING, SYNTHETIC BIOLOGY, AND OTHER POTENTIALLY TRANSFORMATIONAL RESEARCH AREAS*

United States Government Accountability Office

WHY GAO DID THIS STUDY

Scientific and technological innovation contributes to U.S. economic competitiveness and prosperity. Federal agencies support transformational technological advances—those that result in new or significantly enhanced

* This is an edited, reformatted and augmented version of United States Government Accountability Office; Report to the Subcommittee on Research and Technology, Committee on Science, Space, and Technology, House of Representatives, Publication No. GAO-18-656, dated September 2018.

technologies— by, for example, funding research (nearly $70 billion in obligations in fiscal year 2017).

GAO was asked to examine support for research that could lead to transformational technological advances. This report (1) describes federal agencies' and nonfederal entities' support for such research in selected areas, (2) examines federal agencies' coordination on this research, and (3) describes experts' views on considerations for maintaining U.S. competitiveness through such advances. GAO selected quantum computing and synthetic biology as examples of research areas that could lead to transformational technological advances. GAO reviewed agency documents and interviewed federal officials, subject matter experts, and stakeholders. GAO also worked with the National Academies of Sciences, Engineering, and Medicine to convene a meeting to solicit views from 19 experts selected from government, academia, and industry, among others.

What GAO Recommends

GAO recommends that the agencies leading the interagency quantum computing and synthetic biology groups take steps to fully implement leading collaboration practices. The agencies agreed with GAO's recommendations.

What GAO Found

Multiple federal and nonfederal entities support research for transformational technological advances in the areas of quantum computing—the manipulation of bits of data using the behavior of individual atoms, molecules, or other quantum systems to potentially outperform supercomputers—and synthetic biology—the combination of biology and engineering to create or modify biological systems. GAO found that at least 6 agencies support quantum computing research; at least

10 agencies support synthetic biology research; and nonfederal entities, such as universities and businesses, support research in both areas.

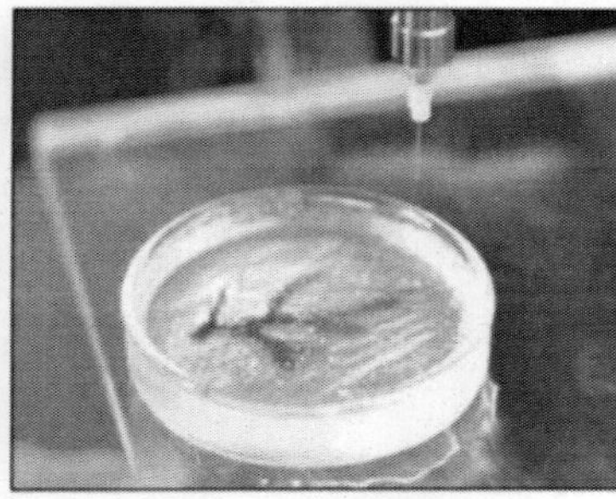

Sources: Massachusetts Institute of Technology Lincoln Laboratory (left) and Adam Feinberg, Carnegie Mellon University (right). GAO-18-656.

Quantum Computing Device (Left) and 3D Bioprinted Coronary Artery (Right).

Agency officials said they coordinate on quantum computing and synthetic biology through efforts such as conferences and interagency groups, but GAO found that certain new efforts have not fully implemented selected leading collaboration practices. The quantum computing group, co-chaired by officials from 4 agencies, and the synthetic biology group, led by the National Science Foundation, have taken initial steps to implement some leading practices GAO identified that can enhance and sustain interagency collaboration. For example, both groups agreed to coordinate their research, and participating agencies documented agreement with the quantum computing group's purpose through a charter. However, the groups have not fully implemented other practices, such as agreeing on roles and responsibilities and identifying common outcomes, that could help ensure they effectively marshal agencies' efforts to maintain U.S. competitiveness in quantum computing and synthetic biology.

Experts identified considerations for maintaining U.S. competitiveness through transformational technological advances. The considerations broadly address federal and nonfederal entities' roles in supporting such advances and include:

- developing a strategic approach using consortia or other mechanisms to bring together potential partners;
- fostering an environment in which information is shared among researchers while also considering the risks of information sharing;
- focusing on technology development and commercialization, for example, by providing support across multiple stages of technology innovation; and
- strengthening the science and technology workforce through training, recruiting, and retaining talent.

ABBREVIATIONS

ARPA-E	Advanced Research Projects Agency-Energy
BER	Biological and Environmental Research
BRAIN Initiative	Brain Research through Advancing Innovative Neurotechnologies Initiative
COMPETES 2007	America Creating Opportunities to Meaningfully Promote Excellence in Technology, Education, and Science Act
CRISPR	clustered regularly interspaced short palindromic repeats
DARPA	Defense Advanced Research Projects Agency
DHS	Department of Homeland Security
DNA	deoxyribonucleic acid
DOD	Department of Defense
DOE	Department of Energy
EPA	Environmental Protection Agency
EPSCoR	Established Program to Stimulate Competitive Research
HHS	Department of Health and Human Services
IARPA	Intelligence Advanced Research Projects Activity
iGEM	International Genetically Engineered Machine
JQI	Joint Quantum Institute

MIT	Massachusetts Institute of Technology
NASA	National Aeronautics and Space Administration
NIH	National Institutes of Health
NIST	National Institute of Standards and Technology
NSCI	National Strategic Computing Initiative
NSF	National Science Foundation
NSTC	National Science and Technology Council
ODNI	Office of the Director of National Intelligence
OSTP	Office of Science and Technology Policy
"qubits"	quantum bits
QIS	quantum information science
QIS Subcommittee	Subcommittee on Quantum Information Science
QIS working group	Interagency Working Group on Quantum Information Science
RNA	ribonucleic acid
SEMATECH	Semiconductor Manufacturing Technology consortium
STEM	science, technology, engineering and math
USDA	U.S. Department of Agriculture

September 26, 2018

The Honorable Barbara Comstock
Chairwoman
The Honorable Daniel Lipinski
Ranking Member
Subcommittee on Research and Technology
Committee on Science, Space, and Technology
House of Representatives

Federal support for research and development can accelerate innovation, drive technological advances, and promote U.S. competitiveness in the global economy. For example, federally supported research led to the introduction of global positioning systems and touchscreen technologies that contributed to the development of smartphones, creating new industries and significant economic growth.[1] In fiscal year 2017, the federal government obligated nearly $70 billion for research, according to the National Science Foundation (NSF).[2] While some research leads to incremental changes in a scientific field, other research can yield disruptive or transformational advances. These advances are transformational because they result in new technologies or significantly enhanced capabilities in existing technologies.

The United States is considered a world leader in many science and technology areas, but other countries, such as China, are also making considerable investments in research. Increased competition from these countries has led some experts and others to express concern that the United States may be losing its competitive advantage. In January 2018, NSF's National Science Board reported that the United States' overall global share of research and development spending is declining relative to other countries.[3] Moreover, other reports indicate that the United States is losing ground in certain technologies, such as intense ultrafast lasers, which may have applications in manufacturing, medicine, and national security.[4]

---

[1] Peter L. Singer, *Federally Supported Innovations: 22 Examples of Major Technology Advances That Stem From Federal Research Support*, (Washington, D.C.: The Information Technology and Innovation Foundation, February 2014).

[2] National Science Foundation, *Survey of Federal Funds for Research and Development Fiscal Years 2015-2017,* January 10, 2018, https://ncsesdata.nsf.gov/fedfunds/2015/index.html. An obligation is a definite commitment that creates a legal liability of the government for the payment of goods and services ordered or received, or a legal duty on the part of the United States that could mature into a legal liability by virtue of actions on the part of the other party beyond the control of the United States. GAO, *A Glossary of Terms Used in the Federal Budget Process*, GAO-05-734SP (Washington, D.C.: September 2005).

[3] The National Science Board also reported that the European Union's global share of research and development spending, overall, declined. National Science Foundation, National Science Board, *Science and Engineering Indicators 2018*, NSB-2018-1 (Alexandria, VA: 2018).

[4] Ultrafast lasers concentrate their energy in a very short-duration pulse (less than a trillionth of a second), which can be focused on a small area to reach the highest peak power and intensity. The National Academies of Science, Engineering, and Medicine, *Opportunities in Intense Ultrafast Lasers: Reaching for the Brightest Light* (Washington, D.C.: National Academies Press, 2017).

Since 2007, a series of laws has built on prior federal efforts to invest in innovation through research and development and to improve the United States' competitiveness.[5] In 2007, the America Creating Opportunities to Meaningfully Promote Excellence in Technology, Education, and Science Act (COMPETES 2007) established the Advanced Research Projects Agency-Energy (ARPA-E) in the Department of Energy (DOE) to overcome long-term and high-risk technological barriers in the development of energy technologies.[6] It also, among other things, authorized programs in the Department of Education and NSF to train teachers in science, technology, engineering and math (STEM) fields. In 2011, the COMPETES Reauthorization Act of 2010 reauthorized parts of COMPETES 2007, established several new programs, and authorized additional funding for STEM education.[7] In 2017, the American Innovation and Competitiveness Act of 2017 reauthorized some parts of the COMPETES acts.[8] The three acts also provided direction to the Office of Science and Technology Policy (OSTP), whose responsibilities include coordinating federal research programs, advising the President on scientific and technological considerations with regard to federal budgets, and advising the President of scientific and technological considerations involved in areas of national concern.

However, continued concerns related to the federal debt and federal budget deficits are forcing difficult decisions about how much the federal government should spend on research, as well as what types of research it should fund. These concerns overlie a long-standing debate about the federal government's role in supporting science and technology research

---

[5] Legislation prior to 2007 includes, for example, the Stevenson-Wydler Technology Innovation Act of 1980, which was enacted for the purpose of improving the economic, environmental, and social well-being of the United States by, among other things, stimulating improved utilization of federally funded technology developments by nonfederal entities. Pub. L. No. 96-480, 94 Stat. 2311 (1980) (codified as amended at 15 U.S.C. §§ 3701-3715, 3719-3723). Another law enacted in 1980, commonly referred to as the Bayh-Dole Act, allowed not-for-profit corporations, including universities, and small businesses to retain title to their federally funded inventions. Patent and Trademark Law Amendments Act, Pub. L. No. 96-517, 94 Stat. 3015 (1980) (codified as amended at 35 U.S.C. §§ 200-211).

[6] Pub. L. No. 110-69, § 5012(b), 121 Stat. 572, 621 (2007) (codified as amended at 42 U.S.C. § 16538(b)).

[7] Pub. L. No. 111-358, 124 Stat. 3982 (2011).

[8] Pub. L. No. 114-329, 130 Stat. 2969 (2017).

and which roles should be left to the private sector, academia, or other nonfederal entities. We have reported that the government often funds research at the early stages of technology development, whereas industry typically supports the final stages of technology development.[9] As a result, U.S. innovators may find it difficult to obtain public funding or private investment during the middle stages of innovation. We have also reported that substantial amounts of funding or investment are needed to support technology maturation through the middle stages of innovation and that high costs can be a barrier to technology commercialization, especially for small and medium-sized U.S. enterprises. Long-standing disagreements among stakeholders about the appropriateness of the federal government taking an active role in supporting technological advances with ambiguous applications has led to the creation, and sometimes the dissolution, of programs such as the Advanced Technology Program and the Technology Innovation Program. These two programs were established to help U.S. businesses and organizations support, promote, and accelerate innovation through high-risk, high-reward research in areas of critical national need.[10] Different perspectives on the appropriate federal role may have also led to the proposals in the President's Budget for fiscal years 2018 and 2019 to eliminate ARPA-E.

You asked us to examine federal and nonfederal support for research that could accelerate innovation and advance U.S. competitiveness. This report (1) describes federal agencies' and nonfederal entities' support for

---

[9] GAO, *Nanomanufacturing and U.S. Competitiveness: Challenges and Opportunities,* GAO-14-618T (Washington, D.C.: May 20, 2014).

[10] The National Institute of Standards and Technology administered the Advanced Technology Program from fiscal year 1991 through fiscal year 2007 and the Technology Innovation Program from fiscal year 2008 through fiscal year 2010. The Advanced Technology Program was to address long-term, high-risk areas of technological research and development that were not otherwise being adequately developed by the private sector but were likely to yield important benefits to the nation. American Technology Preeminence Act of 1991, Pub. L. No. 102-245, § 201(b), 106 Stat. 7, 15 (1992). In 2007, the Advanced Technology Program was repealed and replaced with the Technology Innovation Program, which was established for the purpose of assisting U.S. businesses and institutions of higher education, as well as other organizations, to support, promote, and accelerate innovation in the United States through high-risk, high-reward research in areas of critical national need. COMPETES 2007, Pub. L. No. 110-69, § 3012, 121 Stat. 572, 593 (2007). The Technology Innovation Program was repealed in 2017 by the American Innovation and Competitiveness Act. Pub. L. No. 114-329, § 205(a)(1), 130 Stat. 2969, 3000 (2017).

research for transformational technological advances in selected areas, (2) examines federal agencies' coordination on this research, and (3) provides experts' views on considerations for maintaining U.S. competitiveness through transformational technological advances.

To address these objectives, we selected quantum computing (a sub-area of quantum information science) and synthetic biology as examples of areas of research that could lead to transformational technological advances. We selected these two areas based on several factors, including that they (1) represent enabling or platform technologies, (2) are supported by a mix of federal agencies and nonfederal entities, and (3) represent areas of congressional interest in which we have not recently conducted work.[11] For the purposes of our report, we defined quantum computing as computing in which bits of data are manipulated by using the behavior of atoms, molecules, or other quantum systems, with the potential to carry out extremely complicated calculations for specific problems that can outperform conventional supercomputers. We defined synthetic biology as the intersection of biology and engineering that focuses on the modification or creation of novel biological systems for useful purposes.

To describe federal agencies' and nonfederal entities' support for research that could lead to transformational technological advances in quantum computing or synthetic biology, we focused on federal and nonfederal efforts in fiscal year 2016 through the second quarter of fiscal year 2018. We reviewed agency documentation, relevant literature, and our prior work related to federal research efforts. We also interviewed officials from federal agencies that support quantum computing or synthetic biology research, as well as subject matter experts in the areas of quantum computing, synthetic biology, or federal research more broadly from industry, academia, nonprofit organizations, and professional associations. We included 10 agencies in our review: Department of Commerce,

[11] Enabling technology is equipment and/or methodology that, alone or in combination with associated technologies, provides the means to generate giant leaps in performance and user capabilities. A platform technology represents technologies that are used as a base on which other applications, processes, or techniques are developed. These technologies function as innovation catalysts and facilitate the development of follow-on technologies with applications for various industries.

Department of Defense (DOD), Environmental Protection Agency (EPA), DOE, Department of Homeland Security (DHS), Department of Health and Human Services (HHS), National Aeronautics and Space Administration (NASA), NSF, Office of the Director of National Intelligence (ODNI), and the U.S. Department of Agriculture (USDA).[12] We did not seek to develop comprehensive information on federal agencies' and nonfederal entities' efforts to support research in quantum computing and synthetic biology. As a result, federal agencies and nonfederal entities could have efforts in these two areas that we do not discuss in our report.

To examine federal agencies' coordination on quantum computing and synthetic biology research, we identified coordination efforts that took place in fiscal year 2016 through the second quarter of fiscal year 2018 during our review of agency documentation and interviews with federal officials, including OSTP officials. We then compared these efforts with selected leading practices for enhancing and sustaining collaboration.[13] We selected six of the eight leading practices based on their relevance to the operations of the interagency coordination efforts we identified.[14] In this report, and in our past work, we define collaboration as any joint activity that is intended to produce more public value than could be produced when organizations act alone.[15]

To provide experts' views on considerations for maintaining U.S. competitiveness through transformational technological advances, we convened a meeting of 19 experts in October 2017, with the assistance of the National Academies of Sciences, Engineering, and Medicine.[16] The experts included current and former federal officials and subject matter

[12] We refer to departments and agencies, collectively, as agencies in our report. Also, as part of our efforts to collect information from DOE, we collected information from the National Nuclear Security Administration.

[13] GAO, *Results-Oriented Government: Practices That Can Help Enhance and Sustain Collaboration among Federal Agencies*, GAO-06-15 (Washington, D.C.: Oct. 21, 2005).

[14] We excluded from our review two leading practices related to reinforcing agency accountability and individual accountability for collaborative efforts.

[15] We also refer to coordination as collaboration in our work.

[16] We planned and convened this meeting of experts with the assistance of the National Academies to better ensure a breadth of expertise; however, we were responsible for all final decisions regarding meeting substance and expert participation. Any conclusions and recommendations in our reports are solely our own.

experts from industry, academia, nonprofit organizations, and professional associations. About half of the experts were subject matter experts in quantum computing or synthetic biology, while the other half were experts with broader perspectives on the role of federal and nonfederal entities in supporting research for transformational technological advances. We worked with the National Academies to select experts with a range of viewpoints. We used a transcript of the meeting in analyzing information obtained from these experts. See appendix I for more detailed information on the scope and methods of our review and appendix II for a list of the experts who participated in the meeting we convened.

We conducted this performance audit from November 2016 to September 2018 in accordance with generally accepted government auditing standards. Those standards require that we plan and perform the audit to obtain sufficient, appropriate evidence to provide a reasonable basis for our findings and conclusions based on our audit objectives. We believe that the evidence obtained provides a reasonable basis for our findings and conclusions based on our audit objectives.

BACKGROUND

This section provides information on translating research into new products or services, the federal government's role in supporting research, and OSTP's role in fostering collaboration among the various entities. It also provides information on the areas of quantum computing and synthetic biology.

Translating Research into New Products or Services

Technological innovation involves not only creating new ideas but also translating those ideas into a new product or service.[17] Innovation, and the

[17] National Research Council of the National Academies, *Building the Illinois Innovation Economy: Summary of a Symposium* (Washington, D.C.: National Academies Press, 2013).

research driving it, is inherently risky because the likelihood that research can be translated into a product or service and the ultimate value of that product or service are unknown. Because of this risk and the long time frames sometimes associated with technology development, there can be a gap in funding and investment support that makes it challenging to translate research into commercialized products or services. While government and universities often support early-stage research and industry tends to support later stages of development, there may be a gap during the middle stages of innovation during which innovators may have difficulty finding financial support, as illustrated in figure 1 (see app. III for a printable version).[18]

The linear, or pipeline, model of innovation presents innovation as a succession of outputs that transfer to the next level as inputs. The starting point in the pipeline model is basic research. Knowledge created through basic research transitions to the next stage of applied research then to development and, finally, commercialization. Under this model, innovation takes place in distinct and sequential phases.

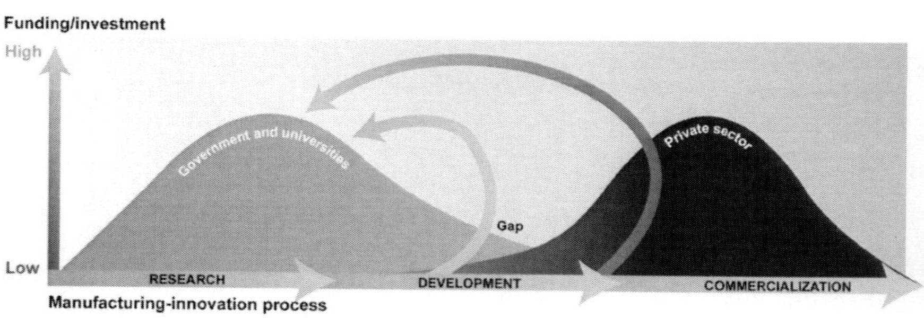

Source: GAO. | GAO-18-656.

Interactive Graphic

Instructions: Online, roll your mouse over each arrow in the figure for more information.

For printable version, see Appendix III, page 75.

Figure 1. Funding/Investment Gap in the Manufacturing-Innovation Process.

---

[18] GAO, *Nanomanufacturing: Emergence and Implications for U.S. Competitiveness, the Environment, and Human Health*, GAO-14-181SP (Washington, D.C.: Jan. 31, 2014).

Critics of the pipeline model have noted that innovation is actually cyclical because the development of knowledge involves feedback and interaction at these different stages of the cycle.[19] Alternative innovation models include the following:

- **Extended pipeline model.** Under this model certain research and development organizations support the entire technology development process, from basic research to initial commercialization.[20] Unlike the pipeline model, in which the government's support is disconnected from the rest of the innovation ecosystem, under the extended pipeline model the government's role is deeply connected to the rest of the system. Under this model, federal entities such as DOD support the evolution of technologies, including electronics, computing, and the internet, across all stages of innovation.
- **Induced innovation model.**[21] Innovation that follows this model is more industry-led because the parties involved have a market niche that the research needs to meet. Research under this model is more likely to lead to incremental advances because it is conducted in response to market demand.[22]
- **Manufacturing-led model.** Under this model,[23] innovation is pursued with the main objective of manufacturing. This model describes innovations in production technologies, processes, and products that emerge from the manufacturing process. The production process is supplemented by applied research and

---

[19] William B. Bonvillian and Charles Weiss, *Technological Innovation in Legacy Sectors* (New York: Oxford University Press, 2015), 23-30. Department of Commerce, *Between Invention and Innovation: An Analysis of Funding for Early-Stage Technology Development* (Gaithersburg, MD: Nov. 2002).

[20] William B. Bonvillian and Peter L. Singer, *Advanced Manufacturing: The New American Innovation Policies*, (Cambridge, MA: The MIT Press, 2017) 8.

[21] This model is also sometimes referred to as a technology demand or technology pull model. See Vernon Ruttan, *Technology Growth and Development: An Induced Innovation Perspective* (New York: Oxford University Press 2001) 7.

[22] Bonvillian and Singer, *Advanced Manufacturing: The New American Innovation Policies*, 7. Bonvillian and Weiss, *Technological Innovation in Legacy Sectors,* 25. Ruttan, *Technology Growth and Development*, 7.

[23] Bonvillian and Singer, *Advanced Manufacturing*, 8.

development. It is typically industry-led but may have strong government support, particularly in countries such as Germany, Japan and China whose economies are organized around this model.

Federal Role in Supporting Research

While the different innovation models receive various levels of federal support, examining the organization of federal agencies in support of innovation is complex because of the decentralized nature of the federal research system. More than 25 federal agencies support intramural or extramural research, and these agencies may play different roles in supporting research that may lead to potentially transformational technologies. For example, NSF supports basic research that is in keeping with its mission of promoting the progress of science; advancing the national health, prosperity, and welfare; and securing the national defense. DOD supports research in line with its mission to provide the military forces needed to deter war and to protect the United States' security, while DOE supports research in line with its mission to ensure America's security and prosperity by addressing its energy, environmental, and nuclear challenges. Commerce's National Institute of Standards and Technology (NIST) supports research in measurement science, standards, and technology, in keeping with its mission to promote innovation and industrial competitiveness. Other agencies—such as EPA, and HHS's Food and Drug Administration—support research in their capacity as regulatory agencies.

Federal support for research is not only decentralized but also changes over time. Factors such as international conflict, budgetary pressures, and globalization may contribute to shifts in U.S. science and technology policy.[24] In times of war, federal support for research has increased in part because of the view that America's military survival might depend on

---

[24] Donald E. Stokes, *Pasteur's Quadrant: Basic Science and Technological Innovation* (Washington, D.C.: Brookings Institution Press, 1997), 90-95.

science and technology leadership.[25] Budgetary pressures also affect the federal role in research when such pressures lead to reductions in federal funding for research. Globalization and the associated integration of the world economy may also affect federal science and technology policy. While the United States invests far more resources in research and development than any other country, its rank in research and development intensity has slowly fallen in recent years.[26]

Researchers have said that, in addition to globalization, domestic changes—such as the structure of U.S. companies—present new challenges to commercializing new products and services.[27] For example, in the last few decades, the amount of research produced by industrial laboratories has declined. Further, U.S. companies, particularly small and midsized firms, devote fewer resources to train employees compared to firms from the 1980s.[28] In recognition of the need for a more skilled workforce to enhance U.S. competitiveness, the federal government has increasingly shifted attention to preparing students for careers in STEM fields.[29]

The federal role also changes in response to differing policy views. One policy perspective maintains that the federal role should be to support innovation across the economy. This policy approach has underpinned innovation and economic growth since at least the end of World War II. As we reported previously, another perspective is that the federal role should be to support individual sectors.[30] Critics of the latter perspective argue

---

[25] Stokes, *Pasteur's Quadrant*, 91.

[26] The National Science Board defines research and development intensity as the ratio of research and development expenditures to gross domestic product. In 2018, the National Science Board reported that the United States' rank by the "research and development intensity" indicator fell from 8th in 2009 to 10th in 2011 and to 11th in 2013 and 2015. National Science Board, *Science and Engineering Indicators 2018*.

[27] Suzanne Berger, *Making In America: From Innovation to Market* (Cambridge, MA: The MIT Press, 2013), 15-19.

[28] Berger, *Making In America,* 19. Bonvillian and Singer, *Advanced Manufacturing*, 221- 222.

[29] GAO, *Science, Technology, Engineering, and Mathematics Education: Assessing the Relationship between Education and the Workforce*, GAO-14-374 (Washington, D.C.: May 8, 2014). As part of this effort, many federal agencies administer STEM education programs. In addition to the federal effort, state and local governments, universities and colleges, and the private sector have also developed programs that provide opportunities for students to pursue STEM education and occupations.

[30] GAO-14-181SP.

that the government should not "pick winners and losers" in commercial contexts because it is unlikely that the government will have sufficient information or foresight about an individual firm's or a particular technology's growth potential to select it for special subsidy.[31] This view advocates allocating resources through market mechanisms because such mechanisms are anticipated to result in U.S. investments that are most efficient and best suited to the comparative advantages of the United States.[32] However, the federal government has supported individual sectors from research and development through implementation, most often because of the government's own needs in areas deemed important for national security (e.g., aerospace and defense).[33] In addition, findings of economic market failures have justified other interventions, such as for research, development, and demonstrations in various sectors, including agriculture and energy, and recently, advanced production technologies.[34]

The federal government has partnered with nonfederal entities to translate research into commercialized products to foster economic growth.[35] For example, DOD, through programs such as the Defense Advanced Research Projects Agency (DARPA), has partnered with nonfederal entities to support both early-stage research and later-stage production.[36] Some of these partnerships have led to development of transformational technologies. For example, in the 1970s DOD supported development of a communications network to facilitate information sharing, which is considered the foundation of the modern internet.[37]

---

[31] GAO-14-181SP.

[32] GAO-14-181SP.

[33] GAO-14-181SP. Vernon W. Ruttan, *Is War Necessary for Economic Growth? Military Procurement and Technology Development* (New York: Oxford University Press, 2006) 3-9.

[34] President's Council of Advisors on Science and Technology, *Report to the President on Ensuring American Leadership in Advanced Manufacturing* (Washington, D.C.: June 2011).

[35] President's Council of Advisors on Science and Technology, *Report to the President on Capturing Domestic Competitive Advantage in Advanced Manufacturing* (Washington, D.C.: July 2012); and GAO, *Advanced Manufacturing: Commerce Could Strengthen Collaboration with Other Agencies on Innovation Institutes*, GAO-17-320 (Washington, D.C.: Apr. 6, 2017).

[36] GAO, *Military Acquisitions: DOD Is Taking Steps to Address Challenges Faced by Certain Companies*, GAO-17-644 (Washington, D.C.: July 20, 2017).

[37] GAO-17-644.

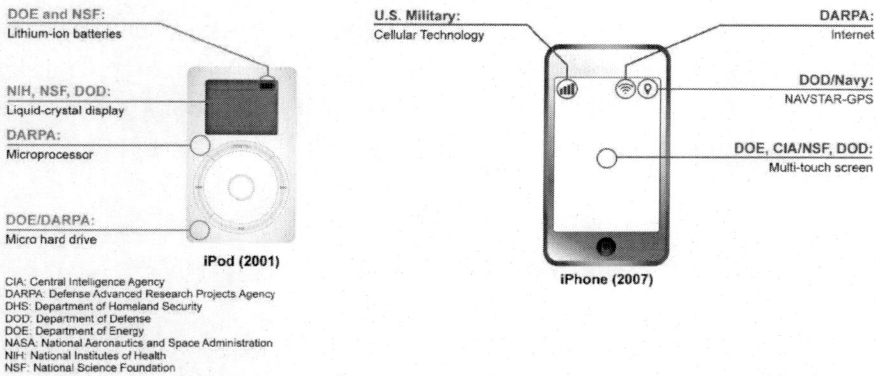

Source: GAO graphic based on information from the Breakthrough Institute, Mariana Mazzucato, National Science Foundation and the Office of Science and Technology Policy. | GAO-18-656.

Note: See Breakthrough Institute, *Where Good Technologies Come From* (Oakland, CA: 2010) and Mariana Mazzucato, *The Entrepreneurial State: Debunking Public vs. Private Sector Myths* (New York, NY: Anthem Press, 2013). According to NIST officials, NIST research in measurements and standards for networking, semiconductors, communications, materials, and cybersecurity contributed to technologies used to develop the iPod and iPhone. Other examples of NIST research contributions to smartphone technologies are in the areas of time and frequency, computing power, and radio communications.

Figure 2. Federally Funded Research That Contributed to Development of Technologies Used to Create the iPod and iPhone.

DOD also funded research in the 1950s on speech recognition and artificial intelligence that commercial companies leveraged in the 1990s and 2000s to develop technologies such as the Speech Interpretation and Recognition Interface, the iPhone assistant.[38] NIST research, such as its critical technical evaluations of speech recognition technologies dating back to the 1980s, also contributed to the development of Speech Interpretation and Recognition Interface, according to NIST officials. Alongside DOE, HHS and NSF, DOD has funded research that led to technologies used to make the first iPod and later the iPhone (see Figure

[38] GAO-17-644.

2).[39] Many federal agencies also support other mechanisms, such as Small Business Innovation Research and Small Business Technology Transfer grants, to stimulate innovation by facilitating interactions among the federal government, private sector, and nonprofit research institutions.

Role of OSTP in Fostering Collaboration

OSTP was established in 1976 to provide advice on the scientific, engineering, and technological aspects of issues that require attention at the highest levels of government.[40] Advances in technology in areas such as quantum computing and synthetic biology have become increasingly interdisciplinary, and OSTP works with agencies across the decentralized federal research system to coordinate activities to support these advances. The National Science and Technology Council (NSTC) is a key component of these efforts and is charged with coordinating science and technology policy across the federal government.[41] One of the NSTC's primary objectives is to establish clear national goals for federal science and technology investments. NSTC organizes its work under six committees, such as the Committee on STEM Education, which is responsible for coordinating federal programs and activities in support of STEM education.[42] In addition to pulling together federal entities, OSTP also

---

[39] Singer, *Federally Supported Innovations*; Mariana Mazzucato, *The Entrepreneurial State: Debunking Public vs. Private Sector Myths*. (New York, NY: Anthem Press, 2013); and Office of Science and Technology Policy, *American Competitiveness Initiative: Leading the World in Innovation* (Washington, D.C.: 2006).

[40] National Science and Technology Policy, Organization, and Priorities Act of 1976, Pub. L. No. 94-282, § 204, 90 Stat. 459, 463 (1976).

[41] The NSTC was established by Executive Order 12881 on November 23, 1993. The council is chaired by the President, and its members include the Vice President, cabinet secretaries and agency heads with significant science and technology responsibilities, and other White House officials. The Assistant to the President for Science and Technology convenes meetings of the NSTC and, in the President's absence, presides over the meetings.

[42] The NSTC's six committees are the Committees on Environment, Homeland and National Security, Science, Science and Technology Enterprise, STEM Education, and Technology. NSTC sometimes organizes efforts on more specific topics under these committees. For example, the National Nanotechnology Initiative, which will be discussed in more detail later in this report, is an effort to bring federal agencies together to enhance understanding

plays a role in pulling together nonfederal entities to help tackle technological issues of importance to the nation. For example, the National Strategic Computing Initiative, created in 2015, is a government collaboration with industry and academia to sustain and enhance U.S. leadership in high-performance computing.[43]

Quantum Computing

Quantum computing has the potential to revolutionize computing by introducing a fundamentally new approach to computing not available with classical computers, which constitute most computers in use today. Classical computers process two different states as 1s and 0s (binary digits) to form "bits" of information that the computer manipulates. Bits can exist in either a 1 or 0 state. These bits may be created using, for example, specific voltage or current levels in a circuit, and there is a limit as to how quickly transistors in classical computers can manipulate these bits to conduct calculations or how many circuit components can be included on a computer chip. While classical computers rely on bits, quantum computers rely on quantum bits ("qubits").[44] Unlike bits, qubits can be in combinations of both a 1 and a 0 at the same time due to quantum superposition. Phenomena such as quantum superposition and quantum entanglement (the ability of two particles to have correlated information, even at a distance) make quantum computers more powerful than even today's most advanced classical supercomputers for solving some complex

---

and control of nanoscale material and is coordinated by the NSTC Committee on Technology's Nanoscale Science, Engineering, and Technology Subcommittee.

[43] Executive Order 13702 established the National Strategic Computing Initiative (NSCI) on July 29, 2015. DOE, DOD, and NSF lead the NSCI effort, and OSTP and the Office of Management and Budget co-chair the NSCI Executive Council. The following agencies also participate in the NSCI: DHS, Federal Bureau of Investigation, Intelligence Advanced Research Projects Activity (IARPA), NASA, National Institutes of Health (NIH), NIST, National Oceanic and Atmospheric Administration, and National Security Agency.

[44] *American Leadership in Quantum Technology, Before the H. Comm. on Science, Space, and Technology,* 115th Cong. 31 (2017) (statement of Carl J. Williams, Acting Director, Physical Measurement Laboratory, National Institute of Standards and Technology Department of Commerce).

problems.[45] This ability to exist in combinations of both states simultaneously allows for the efficient implementation of certain algorithms, resulting in the ability to solve certain types of problems significantly faster than classical computers.[46]

To date, a universal quantum computer is not commercially available.[47] As of 2017, quantum computers contain at most 50 qubits and can perform some small calculations more slowly than classical computers. Among the challenges to building a quantum computer are developing software and hardware. Quantum hardware allows the computer to manipulate qubits by completely isolating quantum processors from outside forces. Quantum computing hardware is at the laboratory prototype stage and is progressing steadily, according to a 2016 federal report.[48] Hardware development efforts include the creation of logical qubits, which use error correction techniques to actively mitigate errors, thus stabilizing the quantum state of the qubit even in the presence of external factors (i.e., noise). Quantum information is extremely fragile and requires special techniques and equipment, such as extreme refrigeration, to maintain the qubit. Other challenges include creating qubits of high quality, packaging them together in a scalable form so they can perform complex calculations in a controllable way, and limiting the errors that can result from heat and electromagnetic radiation. Addressing these challenges may require developing new materials. Stakeholders still consider developing a universal quantum computer a long-term goal. When available, these computers could provide new computational methods and powerful new tools for researchers. Quantum computing has the potential to support significant breakthroughs in medicine, manufacturing, artificial

---

[45] *American Leadership in Quantum Technology* (2017).

[46] An algorithm is a step-by-step procedure for solving a problem or accomplishing some end especially by a computer, according to Merriam-Webster's Collegiate Dictionary.

[47] A universal quantum computer would be able to solve any computational problem, as opposed to a more specialized quantum computer that would have more limited use. According to an industry research laboratory, the universal quantum computer is the most powerful and the hardest to build, posing a number of difficult technical challenges. Current estimates indicate that this machine would comprise more than 100,000 physical qubits.

[48] National Science and Technology Council, *Advancing Quantum Information Science: National Challenges and Opportunities* (Washington, D.C.: July 22, 2016).

intelligence, defense, and improved cybersecurity. However, it may take a decade or more before such technology is ready to be demonstrated at scale.

Synthetic Biology

Synthetic biology represents an intersection of biology and engineering that focuses on the modification or creation of novel biological systems. The current state of synthetic biology is mostly the result of research in biology, engineering, computer science, and information technology dating back to the mid-1900s. Synthetic biology has drawn increasing attention as a potentially transformative platform technology. Whether found in nature or synthesized in a test tube, the building blocks of synthetic biology are assembled to create biological systems. Synthetic biological systems can function in cell-free environments, such as cell extracts, or may be placed into living cells, such as bacteria, which serve as a "chassis." In the short-term, synthetic biology is enhancing understanding of how living organisms work through progress in the ability to design and construct biological parts.

Synthetic biology is already being applied in a variety of fields. Through the creation of novel biological systems, synthetic biology offers potential solutions to many current challenges, such as climate change, energy needs, and global health. For example, synthetic biology may help address global warming through the development of artificial leaf technology, a synthetic version of the photosynthesis process.[49] In the energy sector, synthetic biology is being used to devise more efficient methods of producing biofuels,[50] and in the healthcare sector, synthetic

---

[49] The Royal Academy of Engineering, *Synthetic Biology: Scope Applications and Implications*, (London, United Kingdom: May 2009), 2. Julian David, Janna Olmos and Joanna Kargul, "A Quest for the Artificial Leaf," *The International Journal of Biochemistry and Cell Biology*, 66 (2015) 37-44.

[50] Current methods of producing biofuels result from the production of ethanol from sugars or biodiesel from vegetable oils. These methods have the disadvantage of not being particularly efficient processes that waste much of the organic matter or biomass. See The Royal Academy of Engineering, *Synthetic Biology*.

biology may lead to biosensors that can permanently reside in the body to detect and treat abnormalities such as cancer.[51] Synthetic biology has already resulted in biosensors that can detect arsenic in drinking water.[52] Factors that may support growth in synthetic biology applications include a decline in the cost of deoxyribonucleic acid (DNA) sequencing and increases in genetically engineered crop development, expenditures in research and development by biotechnology and pharmaceutical companies, and demand for synthetic genes.[53] On the other hand, biosafety and bio-security concerns about the potential that synthetic biology could be used for nefarious purposes may restrict the short-term growth of synthetic biology.[54]

MULTIPLE FEDERAL AGENCIES AND NONFEDERAL ENTITIES SUPPORT QUANTUM COMPUTING AND SYNTHETIC BIOLOGY RESEARCH FOR TRANSFORMATIONAL TECHNOLOGICAL ADVANCE

Multiple federal agencies and nonfederal entities support quantum computing and synthetic biology research that could lead to transformational technological advances in many areas of the U.S. economy, including energy, medicine, and national security.[55] We identified 6 agencies that in fiscal year 2016 through the second quarter of

---

[51] The Royal Academy of Engineering, *Synthetic Biology*.

[52] Matthew Charles Edmundson and Louise Horsfall, "Construction of a modular arsenic-resistance operon in E. coli and the production of arsenic nanoparticles," *Frontiers in Bioengineering and Biotechnology, 3, 160* (2015).

[53] Crystal Market Research, *Synthetic Biology Market by Product and Application: Global Industry Analysis and Forecast to 2025* (New York, NY: NASDAQ OMX Corporate Solutions, Inc., November 2017).

[54] The National Academies of Sciences, Engineering, and Medicine, *Biodefense in the Age of Synthetic Biology*, (Washington, D.C.: The National Academies Press: 2018); and *Synthetic Biology Market*.

[55] There may be additional federal agencies that conduct research in quantum computing and synthetic biology that we do not discuss in this report. Additionally the federal agencies we discuss in this report may have applications in areas outside of those we present in this report.

fiscal year 2018 supported quantum computing research to advance foundational understanding of quantum computing or to develop related hardware and software. We found that 4 of the 6 agencies reported a combined total of at least $23.4 million in obligations to support quantum computing research in fiscal year 2017.[56] Similarly, we identified 10 agencies that, during the timeframe we reviewed, supported synthetic biology research to advance foundational understanding of synthetic biology or knowledge of how to apply it in bioengineering, national security, and biofuels development. We found that 6 of the 10 agencies reported a combined total of at least $211.2 million in obligations to support synthetic biology research in fiscal year 2017.[57] We also identified a variety of nonfederal entities, such as universities and private companies, that conduct research in quantum computing and synthetic biology.

Six Agencies Support Research in Quantum Computing

In fiscal year 2017, 6 agencies—DOD, DOE, ODNI, NASA, Commerce's NIST, and NSF—supported quantum computing research, and 4 of these 6 agencies reported a combined total of at least $23.4 million in obligations toward those efforts. Agency officials, stakeholders, and experts we interviewed told us they expect quantum computers could lead to transformational advances in national security technologies or in technology areas that rely heavily on simulation, such as machine learning for defense capabilities, pharmaceuticals, and materials science for advanced manufacturing.[58] However, there is still uncertainty surrounding the specific applications of quantum computing. Agency officials, stakeholders, and experts told us that they anticipate that quantum computing applications may include large number factoring, optimization

[56] DOD and ODNI did not provide data on obligations.

[57] DHS reported that it does not have any obligations related to synthetic biology research. We were not able to collect data on obligations from NIH because the system that NIH uses to track funding of specific scientific topics does not have a category for synthetic biology. Also, ODNI and USDA did not provide data on obligations.

[58] We define advanced manufacturing as manufacturing that uses innovative technology to improve products or processes.

of certain tasks, and simulation of other quantum systems. Accordingly, agencies' quantum computing efforts included research to advance foundational understanding of quantum information science as well as research to develop the hardware and software needed to build a universal quantum computer.

Foundational Understanding of Quantum Information Science

Federally supported research in areas including the physical sciences, mathematics, and engineering can help advance foundational understanding of quantum information science needed to build a universal quantum computer. Among the 6 agencies that support quantum computing research, examples of agencies' efforts to support research to advance foundational understanding of quantum science include:

- NIST supports foundational quantum research at its laboratories and at several university-based research centers. At one of these centers, the Joint Quantum Institute (JQI) at the University of Maryland, researchers are investigating quantum computing architectures and developing methods to control quantum effects that can be exploited to process information in new ways.[59] Additionally, NIST supports JILA, an institute at the University of Colorado, where scientists explore fundamental questions related to quantum information and atomic and molecular physics.[60]
- NSF supports foundational quantum computing research as part of a broader portfolio of research on quantum phenomena in the Directorates for Mathematical and Physical Sciences, Engineering, and Computer and Information Science and Engineering. For example, under its Physics Division's Quantum Information

[59] The Joint Quantum Institute (JQI) was established in 2006 as a cooperative research institute between the University of Maryland and NIST to facilitate the interchange of ideas among scientists working in the fields of atomic physics, condensed matter, and quantum information. NIST also supports the Joint Center for Quantum Information and Computer Science, which was established in 2014 at the University of Maryland as a complement to the JQI.

[60] In 1995, the Joint Institute for Laboratory Astrophysics changed its name to JILA.

Science and Revolutionary Computing program, NSF supports theoretical and experimental research on quantum-based computing paradigms, information, transmission, and manipulation.[61] Also, the NSF Physics Division's Physics Frontiers Centers program supports university-based centers and institutes in enabling transformational advances through interdisciplinary research across different areas of focus. One of the Physics Frontier Centers that NSF supports is located at the JQI; this center supports research that focuses on studying the controlling and monitoring of quantum phenomena to support quantum engineering. A second Physics Frontier Center is at JILA. Both the JQI and JILA represent partnerships between the NSF and NIST.

- DOE's Office of Science supports foundational quantum computing research as part of its Advanced Scientific Computing Research program, which focuses on discovering, developing, and deploying computational and networking capabilities to analyze, model, simulate, and predict complex phenomena important to DOE and the advancement of science. The program's efforts include partnering with other Office of Science program offices to support research aimed at understanding how future computing technologies, including those based on quantum information science, could impact DOE's mission.[62]

---

[61] According to testimony by the NSF Assistant Director for Computer and Information Science and Engineering before the House Committee on Science, Space, and Technology's Subcommittees on Research and Technology and Energy in October 2017, this program was established in fiscal year 2005 as the Quantum Information Science and Revolutionary Computing program. *American Leadership in Quantum Technology*, Before the Subcommittee on Research and Technology and the Subcommittee on Energy, 115th Congress (2017) (statement of Dr. James Kurose, Assistant Director for Computer Science and Engineering, National Science Foundation).

[62] The research supported by DOE's Advanced Scientific Computing Research program follows a report that summarized a February 2015 DOE-led workshop on quantum computing for science. According to the report, high-fidelity modeling and simulation of physical systems is critical for DOE to address some of the most challenging problems in energy, the environment, and national security, and DOE has a long history of advancing computation and increasing the capabilities for its high-performance computing systems to address these challenges. However, the report noted that constraints on increasing improvements to current computing approaches are leading to consideration of alternative computing

Joint Quantum Institute (JQI)

The JQI is a research partnership between the National Institute of Standards and Technology and the University of Maryland, with the support and participation of the Laboratory for Physical Sciences. JQI was created in 2006 to pursue theoretical and experimental studies of quantum physics in the context of information science and technology. Among other objectives, JQI conducts fundamental research on the engineering and control of systems based on quantum mechanics, which describes the behavior of matter and energy at the smallest physical scales. One attribute of quantum physics is that certain properties of a particle, such as its momentum and position, are not fixed; instead these properties follow probability distributions that describe the likelihood a property may be a particular value. Researchers have also discovered that the quantum states of two separate objects, like two atoms, can be entangled such that the state of one object is correlated with the other. This entanglement makes it possible to move quantum information from one place to another. The phenomena that occur at the quantum scale have the potential to affect disparate economic sectors and could lead to improvements in computing and materials science, among others. For example, researchers at JQI have devised a new chip that generates and steers single photons, which could allow researchers to systematically assemble pathways for single photons and enable new types of optical devices.

An illustration of a photonic chip created by JQI researchers.

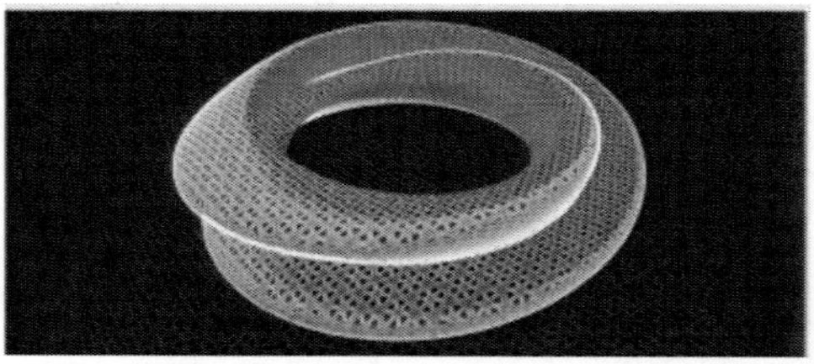

Sources: JQI (text); and Emily Edwards, JQI (image). | GAO-18-656.

approaches, including quantum computing. The February 2015 workshop was organized to assess the viability of quantum computing technologies to meet the computational requirements of DOE's science and energy mission, and to identify the potential impact of quantum technologies. The report found there is great potential for quantum computing to impact DOE's mission; however, additional research in the fundamentals of computer science and mathematics is needed to mature quantum computing technologies into viable solutions for DOE. See Department of Energy, *ASCR Report on Quantum Computing for Science,* (Washington, D.C.: February 17-18, 2015).

- NASA's Quantum Artificial Intelligence Laboratory—a collaborative effort with Google and the Universities Space Research Association— supports foundational research to maximize utilization of emerging quantum hardware. This work involves analytical and experimental research on the mechanisms underlying quantum computing, including, for example, researching quantum entanglement and measurement-based quantum computation. NASA also supports university-based quantum computing research through programs such as the Established Program to Stimulate Competitive Research (EPSCoR).[63]

Hardware Development

Agency officials and experts said that a number of unresolved technical challenges exist related to the hardware necessary to build a quantum computer, including the materials from which to build qubits. Among the 6 agencies that support quantum computing research, examples of efforts to support research to develop the hardware for building a quantum computer include the following:

- DOD supports research related to quantum computing, as part of broader portfolios of research across the department. For example, as part of DOD's Applied Research for the Advancement of Science and Technology Priorities program, the Office of the Secretary of Defense administers the Quantum Science and

---

[63] According to a Congressional Research Service report, NSF established the EPSCoR program—originally named the Experimental Program to Stimulate Competitive Research—in 1978 to address congressional concerns about an "undue concentration" of federal research and development funding in certain states. The program is designed to help institutions in eligible states build infrastructure, research capabilities, and training and human resource capacities to enable them to compete more successfully for open federal research and development funding awards. In addition to NSF, DOE, NASA, NIH, and USDA support EPSCoR or similar programs. See Congressional Research Service, *Established Program to Stimulate Competitive Research (EPSCoR): Background and Selected Issues*, R44689 (Washington, D.C.: Jan. 12, 2017).

Engineering Program—a cross-cutting effort that has supported research related to technologies for controlling qubit entanglement, among other things. Additionally, DOD supports a research program on Quantum System Sciences at Lincoln Laboratory, a federally funded research and development center operated by the Massachusetts Institute of Technology (MIT). This research encompasses, among other topics, development of quantum-based computation technologies.[64]

- DOE's quantum science research efforts, such as those supported by the Office of Science's Advanced Scientific Computing Research program, includes quantum computing hardware and architecture. After DOE issued its 2015 report on quantum computing for science, the agency held a February 2017 workshop to obtain information from stakeholders on the opportunities and challenges in establishing a quantum testbed to advance quantum computing hardware.[65] Subsequently, DOE issued solicitations in 2017 and 2018 for proposals to support developing quantum testbeds. According to an April 2018 announcement for one of these solicitations, a testbed laboratory will host experimental quantum computing platforms that are not yet ready for commercialization, and will function as a collaborative facility to provide internal and external researchers with access to novel, early-stage quantum computing resources.

---

[64] Federally funded research and development centers are government-funded entities that have long-term relationships with one or more federal agencies to perform research and development and related tasks. These centers are typically entirely federally funded, or nearly so, but they are operated by contractors or other nongovernmental organizations. For past GAO work on these centers, see, for example, GAO, *Federally Funded Research Centers: Agency Reviews of Employee Compensation and Center Performance*, GAO-14-593 (Washington, D.C.: Aug. 11, 2014).

[65] DOE issued a report summarizing the input it received from stakeholders during its February 2017 workshop. See Department of Energy, *ASCR Report on a Quantum Computing Testbed* (Washington, D.C.: February 2017).

Lincoln Laboratory's Quantum Computing Laboratory

The Massachusetts Institute of Technology's Lincoln Laboratory is a federally funded research and development center sponsored by the Department of Defense that researches and develops a broad array of advanced technologies to meet critical national security needs. In the area of quantum information science, researchers with Lincoln Laboratory's Quantum Computing Laboratory are exploring the fundamentally different ways that information can be stored and manipulated through quantum physics. Specifically, Lincoln Laboratory researchers are working to develop and scale up two systems that could comprise the quantum bits, or "qubits" of a quantum computer. In one method, called Josephson junction-based superconducting circuits, Lincoln Laboratory researchers are using cryogenic dilution refrigerators and microwave test and measurement equipment to control and measure superconducting qubits at extremely cold temperatures. In another method, researchers are using cryogenically cooled vacuum systems to house micro-fabricated chips that trap individual strontium and calcium ions, which are manipulated using lasers and other electromagnetic fields. For both methods, researchers are working to scale up systems of qubits to a size large enough to address real computational problems.

Laser light manipulation of trapped ion qubits at Lincoln Laboratory.

Source: Massachusetts Institute of Technology's Lincoln Laboratory. | GAO-18-656.

- NIST's quantum science research efforts include projects within its Physical Measurement Laboratory that are looking at a spectrum of potential quantum computing hardware approaches, such as superconducting circuits or ion trap-based quantum computing, that could provide viable approaches for processing and manipulating quantum information. By working across multiple approaches, NIST has been able to apply different quantum hardware platforms to address computing and metrology problems, including creating one of the most advanced ion trap-based quantum computing platforms. Furthermore, NIST is using its advanced microfabrication facilities to develop a broad array of components that will enable the scaling of different quantum computing hardware platforms.

- ODNI, through the Intelligence Advanced Research Projects Activity's (IARPA) Logical Qubits Program, is supporting research to overcome the limitations of current multi-qubit systems, whereby qubits are impacted by other qubits, environmental factors, and other forces, which can generate errors in quantum computing operations. IARPA's Logical Qubits Program is sponsoring research teams to build qubit structures with reduced susceptibility to these types of problems and has developed a quantum system with between 10 and 20 qubits.
- NSF supports research related to quantum computing hardware as part of a broader portfolio of research under its Computing and Communication Foundations Division, which supports research that explores the foundations of computing and communications devices and their usage, including advancing hardware designs for computers and computational sciences, among other focus areas. For example, under the division's Expeditions in Computing program, which provides financial assistance awards of up to $10 million over 5 years, NSF provided an award for the Enabling Practical-Scale Quantum Computation project in 2018. This project is a multi-institution, university-based effort to build a 100-qubit computer.

Software Development

Agency officials, stakeholders, and experts said one area in which a quantum computer could offer potential benefits over a classical computer is solving optimization problems. However, using a quantum computer for this or other applications requires developing software to, for example, translate algorithms into the steps to manipulate qubits to perform computing operations. Among the six agencies that support quantum computing research, examples of agencies' efforts to support research to develop software necessary to operate a quantum computer include the following:

- DOD's Air Force Research Laboratory issued a multi-year funding opportunity announcement for research on Quantum Computing Sciences with a focus on quantum computing algorithmic implementation and problem solving. Among other potential research topics, the Air Force is seeking research proposals to develop new algorithms to help solve optimization and machine learning problems.
- NASA's Advanced Supercomputing Division provides funding for the Quantum Artificial Intelligence Laboratory. Through this effort, NASA hosts a 2,031-qubit D-Wave 2000 quantum device. NASA researchers are using this system to explore the potential for quantum computers to tackle optimization problems that are difficult or impossible for traditional supercomputers to handle and to explore the software algorithms that would be needed to do so.

Ten Agencies Support Research in Synthetic Biology

In fiscal year 2017, 10 agencies—DOD, DHS, DOE, EPA, HHS, ODNI, NASA, NIST, NSF, and USDA—supported synthetic biology research, and 6 of these agencies reported a combined total of at least $211.2 million in obligations toward those efforts. According to one agency official and experts, although synthetic biology has advanced significantly, foundational understanding is still needed in some key areas, including measurement and tool development. Accordingly, synthetic biology research that federal agencies supported included research to advance foundational understanding of the science, and the application of synthetic biology in specific areas, such as bioengineering, genome editing, national security, and biofuels and bioproduct development.

Foundational Understanding of Synthetic Biology

One agency official and experts said that although synthetic biology research has made advances, additional foundational work is needed to move the area forward. For example, better measurements are needed in the area of synthetic biology and would be particularly useful for genome-editing methods. Among the 10 agencies that support synthetic biology research, officials from NASA, NIST, DOE, HHS (specifically, the National Institutes of Health (NIH)), and NSF said their agencies support research to advance foundational understanding of synthetic biology. Examples of federal efforts in this area include the following:

Genome in a Bottle

The Genome in a Bottle consortium is one of several ongoing collaborations among the National Institute of Standards and Technology, Stanford University, and other partners in the Joint Initiative for Metrology in Biology. The initiative focuses on measurements and standards supporting the newest developments in genomics and synthetic biology. The Genome in a Bottle consortium focuses on genome sequencing, which involves determining the chemical building blocks of deoxyribonucleic acid (DNA) or ribonucleic acid (RNA) and can give insights into the genes carried by an individual and how and when they are activated. Since the completion of the Human Genome Project in 2003 that first sequenced the whole genome of a human, scientists have worked to make whole human genome sequencing faster and less expensive. The consortium aims to develop the tools needed to ensure the accuracy of human genome sequencing. These tools include reference materials, standards, and data to enable the translation of whole human genome sequencing to clinical practice.

Illustration of a chromosome inside a bottle.

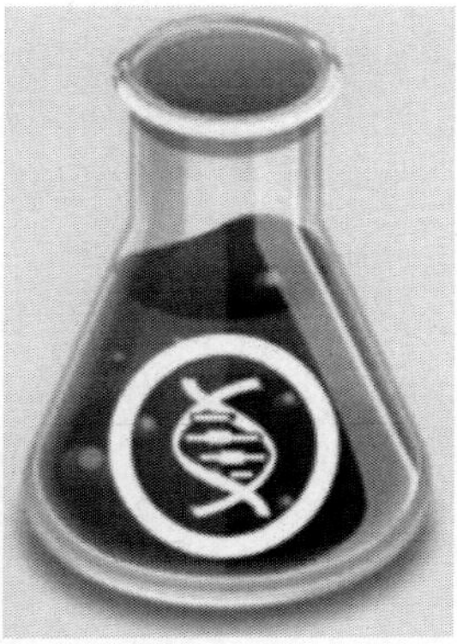

Sources: Joint Initiative for Metrology in Biology, the National Institutes of Health, and GAO (text); Genome in a Bottle consortium (image). | GAO-18-656.

- NIST supports foundational synthetic biology research by developing measurement solutions, serving as a neutral ground for the discussion of underpinning measurements and other manufacturing needs, and leading and contributing to the development of standards. NIST measurement infrastructure includes the development of enabling tools, methods, and protocols; bioinformatics and modeling tools; and documentary standards and reference materials. NIST also leads several consortia to work with measurement stakeholders and partners to accelerate breakthroughs in genomics and synthetic biology. These include NIST's Genome in a Bottle consortium and the Joint Initiative for Metrology in Biology.[66]
- DOE supports foundational research related to synthetic biology as part of a broader portfolio of research under the Biological and Environmental Research (BER) Genomic Science program, which seeks to understand how genomic information is translated to functional capabilities, enabling more confident redesign of

[66] The Joint Institute for Metrology in Biology is a collaboration among NIST, Stanford University, and industry partners to foster the innovation of standards-based measurement to facilitate translation of breakthroughs in genomics and synthetic biology.

microbes and plants for sustainable biofuel production, improved carbon storage, or contaminant bioremediation. Within BER, DOE funds the Joint Genome Institute to produce high-throughput sequencing, a fast method of determining the order of bases of genetic material, synthesis and analysis in support of BER's bioenergy and environmental missions. Research enabled through this user facility includes developing renewable and sustainable sources of biofuels from plant biomass and exploring the biological processes controlling greenhouse gas accumulation in the atmosphere.

- Within HHS, multiple NIH institutes and centers support foundational research involving synthetic biology techniques, including NIH Common Fund support for research to understand and combat antibiotic resistance and National Cancer Institute support for research into new cancer immunotherapy methods. Additionally, the National Institute of Biomedical Imaging and Bioengineering has provided grants to researchers studying or using a multitude of synthetic biology techniques for applications, such as improving stem cell quality for biomedicine.
- NSF funds an estimated $60 million a year in foundational synthetic biology research across several directorates. For example, in 2013, NSF awarded a 5-year, $10 million Expeditions in Computing grant for a multi-university effort led by the California Institute of Technology to enable theoretical investigations in several synthetic biology-related topic areas. In 2016, NSF awarded a second 5-year, $10 million Expeditions in Computing grant for a multi-university effort led by Boston University to support synthetic biology research.

Bioengineering

Applications of synthetic biology in the field of bioengineering may lead to transformational technological advances in areas such as medicine and energy production, according to agency officials, experts, and literature.[67]

Among the 10 agencies that support synthetic biology research, officials from DOD, NASA, EPA, NSF, and NIH said their agencies support synthetic biology research related to bioengineering applications. Examples of federal efforts in this area include the following:

- DOD's DARPA is leveraging biotechnologies, such as synthetic biology, to develop new organisms with unprecedented behaviors and capabilities. DARPA's Living Foundries program funds researchers who are developing the techniques and tools to reprogram living organisms to produce chemicals that would be useful across DOD and to produce a wide variety of chemical compounds that are not easily produced through traditional chemicals. DARPA's goal for the program is to use organisms to make 1,000 molecules and material precursors spanning a wide range of defense-relevant applications. In addition, DOD's Applied Research for the Advancement of Science and Technology Priorities program on Synthetic Biology for Military Environments supports collaborative cross-service synthetic biology research to position DOD to meet unique defense needs and the specific challenges presented by military environments. A key focus of the program is developing several environmental "chassis" microbes that will provide robust performance outside the lab environment. Researchers are also developing applications such as living (self-healing) materials, rugged field-deployable sensors for chemicals and other threats, and microbiome regulation/manipulation for enhanced human performance.
- NASA supports synthetic biology research in the field of bioengineering to increase the capability and reduce the risk of space exploration, and as a tool to advance hypothesis driven investigation. NASA's Advanced Exploration Systems and Game

---

[67] We define bioengineering as the application of principles and techniques from engineering to biological systems.

Changing Development programs are developing on-demand nutrients from microbes engineered to produce targeted nutrients for human consumption as well as examining how to manipulate certain types of bacteria to produce lightweight construction tools and materials.

- EPA employs synthetic biology approaches through its Chemical Safety for Sustainability Research Program, which seeks to develop new prediction techniques, pioneer the use of innovative technologies for chemical toxicity testing, and design tools to advance the management of chemical risks. For example, researchers are developing virtual tissues by building complex computer models for biological development. According to an EPA publication, the models will help reduce dependence on animal study data and provide faster chemical risk assessments.
- NSF's Science and Technology Center Program's Center for Cellular Construction seeks to develop tools to predict, design, and test the impact on cellular function of changes to cells' internal organization. The center will also develop living "bioreactors" that will generate products of commercial value. NSF has funded research into bacterial immunity, which led to the development of clustered regularly interspaced short palindromic repeats (CRISPR)-Cas9—a technology that allows researchers to precisely edit genes.[68]
- Several NIH institutes and centers support research related to bioengineering. For example, NIH's Synthetic Biology for Engineering Applications Funding Opportunity Announcement solicits applications to support research to advance the understanding and application of synthetic biology for human health. In addition, NIH institutes and centers have supported research across various areas, including engineering synthetic receptor systems and genetic controller circuits, engineering

[68] NIH-funded research has also been instrumental to the discovery, understanding, and continued development and application to human health of technologies employed in synthetic biology research such as CRISPR-Cas9 and other gene editing tools.

microbes as therapeutic platforms, and developing enabling technologies for human-machine hybrid tissues.

Gene Editing

The National Institutes of Health (NIH) describes gene editing as a group of technologies that give scientists the ability to add, remove, or alter genetic material at particular locations in the genome. One such technology is known as CRISPR-Cas9, which is short for clustered regularly interspaced short palindromic repeats and CRISPRassociated protein 9. According to NIH, the CRISPR-Cas9 system has generated excitement in the scientific community because it is faster, cheaper, more accurate, and more efficient than other existing gene editing methods. The system was adapted from a naturally occurring gene editing system that helps bacteria defend themselves against viruses by targeting the deoxyribonucleic acid (DNA) of the virus. In the lab, CRISPR-Cas9 allows researchers to cut out a specific sequence of DNA from cells. Once researchers cut out the targeted DNA sequence, they can use other techniques to add or delete genetic material. These genetic changes can cause the edited cells to express new physical traits, such as eye color, or change their disease risk. Gene editing is being applied to research on many diseases; however, according to NIH, there are still significant technical barriers to using gene editing therapies to treat human diseases. Further, the use of gene editing raises a number of ethical concerns.

An illustration of a chromosome unravelling to show the DNA that makes up individual genes.

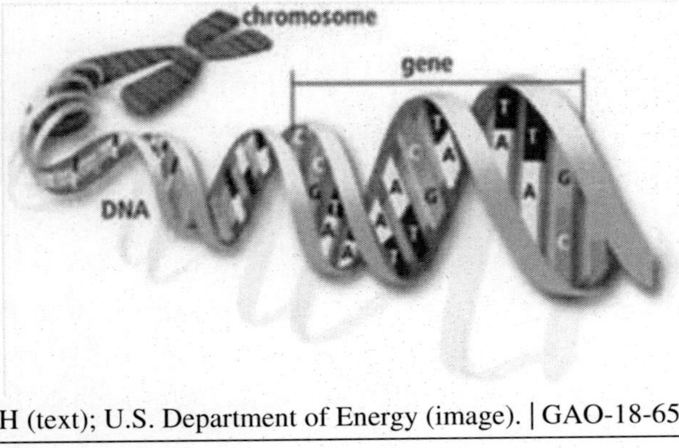

Sources: NIH (text); U.S. Department of Energy (image). | GAO-18-656.

National Security

Application of synthetic biology may support U.S. national security efforts by aiding with monitoring for biological or conventional threats, and strengthening the resilience of soldiers in combat. Among the 10 agencies that support synthetic biology research, officials from DOD, ODNI, and DHS said their agencies support synthetic biology research with potential national security applications. Examples of federal efforts in this area include the following:

- DOD's Office of Naval Research funds research to extend the natural capabilities of living organisms such as microbes and plants to create systems that will provide new naval capabilities, according to the office's website. Office of Naval Research officials told us the office is funding ongoing research related to engineering gut microbes in order to enhance the resilience of service members to deployment stressors, among other things. In addition, DARPA's Safe Genes Project supports force protection and military health and readiness by protecting service members from accidental or intentional misuse of genome-editing technologies. For example, researchers are developing the genetic circuitry and genome-editing machinery for robust, spatial, temporal, and reversible control of genome-editing activity in living systems.
- ODNI supports synthetic biology research through efforts including IARPA's Functional Genomic and Computational Assessment of Threats program, which supports research to protect against critical threats related to pathogens and other biological threats. Researchers aim to develop better approaches and tools for characterization and analysis of biological threats based on gene function.
- DHS's Biological Threat Characterization program and its Biodefense Knowledge Center program support synthetic biology research to understand the risks associated with the technologies

useful for synthetic biology and the harmful pathogens that may be created by those who wish to do harm.[69]

Biofuels and Bioproducts

Synthetic biology is being used to develop cost-effective methods for producing biofuels and bioproducts according to agency officials, experts, and DOE's website.[70] Among the 10 agencies that support synthetic biology research, officials from DOE and USDA said their agencies support synthetic biology research related to biofuels development applications. Examples of federal efforts in this area include the following:

- DOE officials told us that the Office of Energy Efficiency and Renewable Energy's Bioenergy Technologies Office manages the Conversion Program and the Advanced Algal Systems Program, both of which employ synthetic biology techniques to accomplish office goals. Within the Conversion Program, DOE funds the Agile BioFoundry to help develop and transition synthetic biology tools from the laboratory to the biofuels and bioproducts industry. The program accomplishes this through targeted research and development partnerships with industry and academia, as well as by developing integrated synthetic biology tools designed to speed up biomanufacturing. In addition, the office funds the Advanced Algal Systems Program, which supports early-stage applied research to apply synthetic biology approaches to alternative fuels that use algae as their source, among other things. According to a DOE website, this industry has the capability of producing billions of gallons per year of renewable diesel, gasoline, and jet fuel.

---

69 The Biological Threat Characterization program conducts experiments and studies to better understand biological vulnerabilities and hazards. The Biodefense Knowledge Center is a DHS Science and Technology center that produces assessments and tools that help the homeland security community better understand scientific trends that may be exploited by adversaries bent on developing biological weapons or executing biological terrorism.

70 According to the Department of Energy, biofuels are liquid, solid, or gaseous fuel derived from renewable biomass. Biological materials can be used to produce such fuels as biodiesel, ethanol, methanol, methane, and hydrogen. Bioproducts are materials derived from renewable feedstocks. Examples include paper, ethanol, and palm oil.

- USDA, through the Agricultural Research Service, led a collaborative project between federal, industry, and academic researchers to produce a commercial rubber-based tire using the guayule plant, a small shrub native to the United States that has been considered a possible alternative source of natural rubber.

Nonfederal Entities Support Research in Quantum Computing

Nonfederal research to advance quantum computing includes efforts to address existing hardware and software challenges. We identified a variety of nonfederal entities, such as universities and private companies, that have ongoing efforts aimed at building a quantum computer. Stakeholders we spoke to told us that private companies have been increasing their research in quantum computing.

Hardware Development

Academic and industry stakeholders we interviewed described various efforts to develop the hardware needed for a quantum computer. Examples of ongoing efforts include the following:

- Academic researchers at Purdue University partner with Microsoft at Station Q-Purdue to perform a variety of experiments and activities related to building a semiconductor-based quantum computer, including testing different hardware designs.[71]
- Academic researchers from Yale's Quantum Institute are working to develop scalable superconducting devices.
- Researchers at IonQ are working to develop general-purpose quantum information processors using a trapped-ion approach to create a quantum computer that is scalable and that could support a broad array of applications across a variety of industries.

---

[71] Microsoft's partners include the University of California, TU Delft, Niels Bohr Institute, University of Sydney, Purdue University, University of Maryland, ETH Zurich, and other institutions around the world.

- A Google official told us that the company has been working for several years to build a quantum computer through the Quantum Artificial Intelligence Lab. In a March 2018 press release, Google announced its newest 72-qubit quantum computer, called Bristlecone.

Software Development

Academic and industry stakeholders we interviewed described ongoing efforts related to software development. Examples of ongoing efforts include the following:

- An official from Microsoft said the company is working to develop quantum algorithms and software to run on a quantum computer for a given set of problems. Researchers are also currently developing an operating system and various applications that could be run on a quantum device.
- An IBM official told us that, in 2016, the company launched the Quantum Experience, a quantum computing system with five superconducting qubits on the cloud, encouraging students and researchers worldwide to explore quantum computing. Over the past two years, the system's software has been expanded and upgraded for greater functionality and exploration of quantum algorithms to allow researchers around the world to use the system to write more than 80 research publications. MIT and many other universities now use the Quantum Experience in their curricula.

Nonfederal Entities Support Research in Synthetic Biology

We identified a variety of nonfederal entities, such as universities and private companies, that conduct research in synthetic biology to advance foundational understanding and develop new products.

Foundational Understanding of Synthetic Biology

The iGEM Foundation

The International Genetically Engineered Machine (iGEM) Foundation is an independent, non-profit organization dedicated to the advancement of synthetic biology, education and competition, and the development of an open community and collaboration. The foundation does these by fostering an open, cooperative community and friendly competition. The main iGEM program is the iGEM competition, which began in January 2003 as an independent study course at the Massachusetts Institute of Technology in which students developed biological devices to manipulate cells. This course became a summer competition with 5 teams in 2004, grew to 13 teams in 2005, and had expanded to 310 teams by 2017, reaching more than 40 countries. The competition was originally aimed at college students but has expanded to include high school students and others. The iGEM competition gives students the opportunity to push the boundaries of synthetic biology by tackling everyday issues facing the world. Multidisciplinary teams made up of primarily university students work together to design, build, test, and measure a system of their own design using interchangeable biological parts and standard molecular biology techniques. Every year nearly 6,000 people dedicate their summer to iGEM and then come together in the fall to present their work and compete at the annual Jamboree.

A picture of the iGEM logo.

Sources: iGem Foundation (text); iGem Foundation and Justin Knight (image). | GAO-18-656.

Nonfederal stakeholders we interviewed described efforts to advance synthetic biology through activities they initiated as well as through efforts they led with support from federal and industry partners. These efforts include the following:

- The International Genetically Engineered Machine (iGEM) Foundation hosts an annual worldwide synthetic biology competition in Boston, the iGEM Giant Jamboree. The competition attracts teams from around the world (primarily university students) to use standardized genetic parts to address real-world problems in fields including health, medicine, manufacturing, and bioenergy.
- At MIT's Synthetic Biology Center, researchers work with federal and industry partners to advance understanding of synthetic biology for genetic programming, DNA synthesis, and genome design. Researchers at the Synthetic Biology Center seek to create a programming language for living cells that is similar to languages used to program computers and robots.

Development of New Products and Technologies

Industry stakeholders we interviewed described various efforts to develop new products and technologies based on synthetic biology, such as biosensors, new data storage technologies, and bioengineering techniques. These efforts include the following:

- IBM researchers are developing biosensors that may be used for the early detection of cancer. They are also working on understanding and analyzing cardiac, neurological, and mental health conditions.
- Researchers from Microsoft said the company is conducting research related to data storage using synthetic DNA as the information preservation medium. This storage technology uses a process by which custom sequences of synthetic DNA are manufactured to store information.

DNA Storage

To facilitate storing an ever-increasing amount of digital data, researchers from Microsoft, in collaboration with the University of Washington, are studying the use of synthetic deoxyribonucleic acid (DNA) as a means of storing data. According to a Microsoft researcher, this technology uses a process by which custom sequences of synthetic DNA are produced or manufactured to store information. The researcher described three main advantages of storing data in DNA as compared to the current means of storing data, generally magnetic and optical media:

Density. DNA may allow for the storage of up to 1 exabyte (one quintillion bytes) of data per cubic millimeter. In comparison, according to Microsoft, storing similarly large volumes of data in optical discs would occupy significant physical space.

Durability. Data could be stored in DNA for thousands of years, according to one Microsoft researcher.

Relevance. DNA would always remain a relevant storage mechanism, unlike other means of storing digital data (e.g., floppy discs), which becomes outdated as technology advances.

Source: Microsoft. | GAO-18-656.

- Ginkgo Bioworks officials said the company is focused on trying to de-risk supply chains and improve supply chain management through synthetic biology approaches. To that end, the company designs custom enzymes for a variety of customers including companies in a wide range of industries such as food and fragrance companies.
- The Energy Biosciences Institute is a partnership among the University of California, DOE's Lawrence Berkeley National Lab, and the University of Illinois. Researchers at the Energy Biosciences Institute carry out research in the areas of biofuels,

carbon sequestration, and sustainable chemicals productions, among other things.

Agencies Coordinate Research through a Range of Efforts, but Interagency Groups Have Not Fully Implemented Selected Leading Practices

Agency officials we interviewed said they coordinate on quantum computing and synthetic biology research through a range of efforts, but we found that certain efforts are new and that agencies have not fully implemented selected leading practices for collaboration in these efforts. Agency officials told us they use means of coordination ranging from attending ad hoc meetings, such as conferences or workshops, to participating in ongoing interagency groups, such as interagency groups on quantum information science (QIS) and synthetic biology. However, we found that new interagency groups on QIS and synthetic biology have not fully implemented leading practices that can enhance and sustain collaborative efforts.

Agencies Coordinate on Quantum Computing and Synthetic Biology Research Using Efforts That Range from Ad Hoc Meetings to Ongoing Interagency Groups

Agency officials said that they coordinate on quantum computing and synthetic biology research by attending ad hoc meetings, as well as through ongoing efforts such as participating in interagency working groups. The means of coordinating that officials most frequently cited were participating in working groups or attending a conference or workshop. Meetings such as these bring together representatives of different agencies or departments to discuss common problems, exchange information, or develop agreements on issues of mutual interest, as we have reported in the past.[72] Specifically:

- Officials from 4 of the 6 agencies that support quantum computing research said they attended a conference or workshop related to quantum computing at some point from October 2015 through March 2018. For example, NASA and DOE officials participated in a 2017 NASA workshop that brought together experts from NASA research centers, DOE national laboratories, academia, and industry to discuss quantum information science and computation.[73]
- Officials from all 10 agencies that support synthetic biology research cited attendance at a conference, and officials from 7 of these 10 cited workshops as a way in which they coordinated on synthetic biology research from October 2015 through March 2018. For example, officials from DOD, DOE, NIST, and national laboratories attended a 4-day conference in June 2017 to discuss synthetic biology applications in genetic engineering.[74]
- Officials from 7 of the 10 agencies that support synthetic biology research also said they coordinated research with other selected agencies through communities of practice or consortia that meet on an ad hoc basis. For example, NASA officials said they support synthetic biology work through the Space Technology Research Institute in Biomanufacturing, a University of California Berkeley-led consortium of universities.

Officials we interviewed also said they coordinate with one another through ongoing efforts, such as interagency groups. For example, on June 21, 2018, NSTC established the Subcommittee on Quantum Information

---

[72] GAO, *Managing for Results: Key Considerations for Implementing Interagency Collaborative Mechanisms*, GAO-12-1022 (Washington, D.C.: Sept. 27, 2012).

[73] NASA hosted the Quantum Computing for Aerospace and Engineering workshop on Nov. 7-8, 2017, in Suffolk, Virginia.

[74] The Synthetic Biology: Engineering, Evolution and Design conference was held in Vancouver, British Columbia, in June 2017.

Science (QIS Subcommittee) to coordinate quantum computing research.[75] According to its June 2018 charter, the QIS Subcommittee's purpose is to establish and maintain a national agenda in quantum information science and technology, expand U.S. economic and national security, and coordinate federal quantum information science and technology policy and programs. The functions of the QIS Subcommittee include to

- issue and update plan(s) that coordinate(s) federal policy to expand U.S. leadership in quantum information science and technology;
- enable stakeholders to invest effectively in quantum information science and technology and post-quantum application spaces through data gathering, analysis, consultation, planning, convening, and reporting; and
- provide a forum for research and development coordination and collaboration, including sharing expertise and best practices for program management and conducting joint workshops and program reviews.

The QIS Subcommittee is led by co-chairs from NIST, DOE, NSF, and OSTP and includes 9 additional agencies.[76] The QIS Subcommittee met for the first time as an official chartered group on June 28, 2018. The OSTP official serving as a co-chair for the QIS Subcommittee said that the group's first priority will likely be to develop a national approach to QIS research and development.

Officials from 5 of the 6 agencies that support quantum computing research said that prior to the formation of the QIS Subcommittee, they coordinated through the NSTC Interagency Working Group on Quantum Information Science (QIS working group), which was formed in 2014.[77] In

---

[75] The Subcommittee on Quantum Information Science is under NSTC's Committee on Science.

[76] The QIS Subcommittee's member agencies include DOD, DOE, DHS, Department of the Interior, HHS, NASA, NIST, NSF, National Security Agency, ODNI, Department of State, and USDA. Members also include the Office of Management and Budget and OSTP from the Executive Office of the President.

[77] The following agencies were members in the QIS working group: DOD, DOE, DHS, NIST, NSF, National Security Agency, and ODNI. The Office of Management and Budget, OSTP, and National Security Staff from the Executive Office of the President were also members.

July 2016, the QIS working group produced a report, which the agency officials serving as the group's co-chairs told us included its strategic plan for federal QIS research.[78] The July 2016 report identified QIS as a priority for federal coordination and investment as a component of U.S. scientific leadership, national security, and economic competitiveness. The QIS Subcommittee co-chair from OSTP said that the shift from a working group to a subcommittee is a significant elevation that communicates the importance of QIS to the administration.

Agencies also coordinated synthetic biology research through interagency working groups. Officials from NSF and USDA told us that, in December 2017, they formed a new synthetic biology working group that had 7 member agencies as of February 2018.[79] These officials said that the participating agencies saw a need for continued communication and information sharing, and the officials said the group's efforts will increase coordination. Prior to the formation of this new group, 7 of the 10 agencies that support synthetic biology research participated in an NSTC Synthetic Biology Working Group that NSF officials said existed from 2012 to 2013 and was co-chaired by DOD and DOE,[80] according to a 2013 DOE report to Congress that the group produced.[81] According to some officials, the

---

NASA was not a member until 2018, according to agency officials. The QIS working group was preceded by the NSTC Committee on Technology's Subcommittee on Quantum Information Science, which was tasked with developing a vision for federal QIS research. This committee produced the report, *A Federal Vision for Quantum Information Science* (Washington, D.C.: Jan. 5, 2009).

[78] National Science and Technology Council, *Advancing Quantum Information Science*.

[79] Officials from USDA and NSF said that, in addition to their agencies, the working group includes representatives from DOD, DHS, DOE, ODNI, NASA, and OSTP. According to DOD and NSF officials, some member agencies first met in September 2017 to discuss the group. Officials from DHS and EPA also cited the Technical Working Group on Security Risks Associated with Applications Enabled by Advances in Genome Editing and Synthesis Technologies and Other Related Enabling Technologies, chaired by DHS, HHS and OSTP, as a working group through which synthetic biology was discussed in the context of security risks.

[80] According to a DOD official, after the synthetic biology working group ended, officials from several agencies participated in quarterly phone calls from 2014 through 2015 to discuss synthetic biology programs.

[81] Department of Energy, *Synthetic Biology Report to Congress* (Washington, D.C.: July 2013). The interagency Synthetic Biology Working Group was created in response to a Congressional directive to develop a comprehensive plan for federally supported synthetic biology research and development activities. The 7 agencies involved in this working group were USDA, Commerce, DOD, DOE, HHS, NASA, and NSF.

working group ended after it produced this report, which described synthetic biology research and development needs at the time and identified which federal agencies were planning synthetic biology research. The report also discussed the need for communication and coordination among federal agencies that support basic and applied synthetic biology research to build synergies, consider new research and development needs, and evaluate issues as they emerge. According to a senior NSF official we interviewed who was helping lead efforts to establish the new group, one of its first undertakings will be to update the 2013 report to provide a roadmap for agencies' synthetic biology research. However, the official also stated that the participating agencies were still considering the new group's activities.

Agencies Are Coordinating on Quantum Computing and Synthetic Biology through New Interagency Groups, but Have Not Fully Implemented Leading Collaboration Practices

By recently establishing the QIS Subcommittee and a synthetic biology working group, NSTC and federal agencies, respectively, took steps to further coordination on quantum computing and synthetic biology research. However, the new subcommittee and working group have not fully implemented leading practices for collaboration.[82]

We have reported that effective collaboration can help reduce or better manage fragmentation, overlap, and duplication of federal programs.[83] As described above, a number of federal agencies support research related to quantum computing and synthetic biology. In our April 2015 guide to

[82] GAO-06-15 defines collaboration broadly as any joint activity that is intended to produce more public value than could be produced when the agencies act alone.

[83] GAO, *Fragmentation, Overlap, and Duplication: An Evaluation and Management Guide*, GAO-15-49SP (Washington, D.C.: Apr. 14, 2015). GAO-15-49SP defines overlap as when multiple agencies or programs have similar goals, engage in similar activities or strategies to achieve them, or target similar beneficiaries. GAO-15-49SP defines duplication as instances when 2 or more agencies or programs are engaged in the same activities or provide the same services to the same beneficiaries.

evaluating and managing fragmentation, overlap, and duplication, we define fragmentation as those circumstances in which more than one federal agency, or organization within an agency, is involved in the same broad area of national need, and opportunities exist to improve service delivery.[84] This definition applies concerning federal agencies' quantum computing and synthetic biology research, with more than one agency involved in the same broad area of national need. However, as shown in our description above of the agencies' support for research in these two areas, agencies' activities sometimes differ in meaningful ways or leverage the efforts of other agencies.

We examined agencies' efforts to coordinate through interagency groups by selecting six leading practices that we have previously identified can enhance and sustain interagency collaboration:[85]

- **Define and articulate a common outcome.** Effective collaboration requires agencies to define and articulate common outcomes or purposes they are seeking to achieve that are consistent with their respective agencies' goals and missions.
- **Establish mutually reinforcing or joint strategies.** Having mutually reinforcing or joint strategies enables agencies to align activities, core processes, and resources to achieve a common outcome.
- **Identify and address needs by leveraging resources.** Agencies can sustain their collaborative efforts by identifying the human, information technology, physical, and financial resources necessary to achieve identified outcomes.
- **Agree on roles and responsibilities.** By defining and agreeing on roles and responsibilities, including leadership, collaborating agencies can better clarify who will do what, organize their joint and individual efforts, and facilitate decision making.

---

[84] GAO-15-49SP.

[85] GAO-06-15. Because we focused our review on activities of the interagency groups and not on the individual agencies, we excluded from our review two key practices related to reinforcing agency accountability and individual accountability for collaborative efforts.

- **Establish compatible policies, procedures, and other means to operate across agency boundaries.** Agencies can facilitate collaboration by addressing the compatibility of standards, policies, procedures, and data systems that will be used in the collaborative effort.
- **Develop mechanisms to monitor, evaluate, and report on results.** Creating the means to monitor and evaluate collaborative efforts enables agencies to identify areas for improvement.

We identified limitations in agencies' past efforts to coordinate quantum computing and synthetic biology research. In the area of quantum computing, the QIS working group—which preceded the subcommittee—took steps to implement selected leading practices for collaboration, but the group did not fully implement these practices. For example, the QIS working group's July 2016 report broadly identified quantum computing research needs but did not identify common outcomes for agencies' collaborative efforts to advance QIS, including quantum computing. The three senior officials who served as co-chairs of the QIS working group said they were not aware of any federal goals or outcomes for quantum computing research, and DOE officials said that clarifying common goals could help interagency collaboration on quantum computing research. Officials from some agencies cited challenges with collaborating on joint quantum computing projects—for instance, because of variations among agencies on time frames for providing financial assistance.

OSTP officials described the establishment of the QIS Subcommittee as an effort to further previous coordination conducted through the QIS working group. While the QIS Subcommittee has taken initial steps to implement certain leading practices for collaboration, it has not fully implemented the relevant leading collaboration practices we identified. For example, by developing a charter that identifies its high-level purpose and functions and that identifies co-chairs for the group, the QIS Subcommittee has taken initial steps to identify some agencies' roles and to establish means for operating across agency boundaries. Moreover, by having a charter signed by senior officials, the QIS Subcommittee has taken steps to document agencies' agreement to collaborate, which is a key feature of collaborative mechanisms we have identified in our prior work.[86] However, the working group has not defined roles and responsibilities for agencies other than the co-chairs. OSTP officials said that efforts to date have focused on ensuring that all relevant agencies are included in the QIS Subcommittee; the officials also said that agencies' roles and responsibilities for contributing to the subcommittee will evolve.[87] Table 1 provides additional information on the extent to which the QIS Subcommittee has implemented leading practices for collaboration.

With regard to interagency coordination on synthetic biology research, NSF and USDA officials noted that the new synthetic biology working group hoped to, through continued communication and information sharing, address limitations in agencies' coordination that existed prior to its formation. Officials from NSF said the group was needed for communication, information sharing and to leverage resources and DOD officials agreed that the working group was needed. Additionally, one DOD official and one expert said that limited interagency coordination had resulted in lost opportunities to further develop the area of synthetic biology. They also noted that having a national strategy for synthetic biology would be beneficial. Other officials noted that, as in the area of quantum computing, differences in funding timeframes across agencies hinder their ability to coordinate their synthetic biology research. Some of these officials also said such differences make it difficult to develop an integrative roadmap for their research.

Like the QIS Subcommittee, the new synthetic biology working group has taken initial steps to implement some leading practices for interagency collaboration but has not fully implemented the relevant leading collaboration practices we have identified. For example, the group has taken initial steps to identify member agencies' roles by having NSF serve as the lead agency for the first 2 years. However, the group has not identified other member agencies' roles and responsibilities. An NSF

---

[86] GAO-12-1022.

[87] Ensuring that all relevant agencies are included is another key feature of collaborative mechanisms we identified in GAO-12-1022.

official said the new working group had also considered developing a document, such as a charter, to guide its efforts but, as of June 2018, it had not yet decided whether to do so. Table 2 provides additional information on the extent to which the Synthetic Biology Working Group has implemented leading practices for collaboration.

Table 1. Extent of Implementation of Selected Leading Practices for Collaboration by the Quantum Information Science (QIS) Subcommittee, as of June 2018

Leading collaboration practice	Status of QIS Subcommittee efforts
Define and articulate a common outcome	• QIS Subcommittee charter identifies the subcommittee's high-level purpose and functions.
	• Office of Science and Technology Policy (OSTP) officials said the subcommittee has not yet determined whether it will define and articulate common outcomes as part of an effort to develop a long-term plan to address QIS challenges.
Establish mutually reinforcing or joint strategies	• OSTP officials said a first step will be to develop a national approach and plan for QIS research and development.
Identify and address needs by leveraging resources	• OSTP officials said a main goal is to coordinate the agencies' efforts and work through the agencies to achieve goals.
Agree on roles and responsibilities	• QIS Subcommittee charter identifies certain agencies as co-chairs for the group.
	• Other member agencies' roles and responsibilities have not been identified, as of June 2018.
	• OSTP officials said that agencies' roles and responsibilities are still evolving.
Establish compatible policies, procedures, and other means to operate across agency boundaries	• QIS Subcommittee charter outlines the purpose and scope of the subcommittee, its functions and membership, the potential for seeking private sector advice as needed, and the subcommittee's termination date.
	• Specific policies and procedures for collaborating across agency boundaries have not been established, as of June 2018.
Develop mechanisms to monitor, evaluate, and report on results	• QIS Subcommittee charter states that one function of the group includes reporting.a
	• OSTP officials said that no particular mechanism has been identified, as of June 2018.

Source: GAO analysis of information provided by OSTP officials. | GAO-18-656.

[a]According to the QIS Subcommittee charter, this function is: Enable stakeholders to invest effectively in quantum information science and technology and post-quantum application spaces through data gathering, analysis, consultation, planning, convening, and reporting.

Table 2. Extent of Implementation of Selected Leading Practices for Collaboration by the Synthetic Biology Working Group, as of June 2018

Leading collaboration practice	Status of Synthetic Biology Working Group efforts
Define and articulate a common outcome	• National Science Foundation (NSF) officials said the working group will include a roadmap for agencies' synthetic biology research in an update of a 2013 report to Congress on synthetic biology. • NSF officials said there is no time frame for producing the update to the 2013 report.
Establish mutually reinforcing or joint strategies	• NSF officials said the group will update a 2013 report to Congress on synthetic biology that will provide visibility into what is in each agency's portfolio and look for opportunities to collaborate.
Identify and address needs by leveraging resources	• NSF officials said the group will update a 2013 report to Congress on synthetic biology, which will enable agencies to identify where they can collaborate to advance the field.
Agree on roles and responsibilities	• NSF identified as the lead agency for the first 2 years.
	• Roles of member agencies other than NSF have not been identified, as of June 2018.
	• Some member agencies have taken responsibility for drafting sections of a report that will update a 2013 report to Congress on synthetic biology.
	• Other responsibilities have not been identified, as of June 2018.
Establish compatible policies, procedures, and other means to operate across agency boundaries	• No steps taken to establish policies, procedures, or means to operate across agency boundaries, as of June 2018.
Develop mechanisms to monitor, evaluate, and report on results	• No steps taken to develop mechanisms to monitor, evaluate, and report on results, as of June 2018.

Source: GAO analysis of information provided by NSF officials. | GAO-18-656.

As we previously reported, interagency collaborative mechanisms can take many different forms, such as working groups or subcommittees, and the leading practices we identified that help enhance and sustain interagency collaboration can be adapted to help address the specific challenges agencies face.[88] For example, incorporating the leading practices into agencies' collaborative efforts can help address issues associated with potential fragmentation, overlap, and duplication in

[88] GAO-06-15.

instances where multiple agencies have activities in a similar area. The QIS Subcommittee and the synthetic biology working group are mechanisms through which agencies can address limitations in past interagency coordination on quantum computing and synthetic biology. However, as of July 2018, the subcommittee and working group were still new and have had limited time to fully implement the leading practices we have identified. As the subcommittee and the working group move forward, by taking steps to fully implement these leading practices, member agencies could better marshal their collective efforts to support research in the areas of quantum computing and synthetic biology and help maintain U.S. competitiveness through transformational technological advances.

EXPERTS IDENTIFIED KEY CONSIDERATIONS FOR MAINTAINING U.S. COMPETITIVENESS THROUGH TRANSFORMATIONAL TECHNOLOGICAL ADVANCES

Experts who participated in the meeting we convened with the assistance of the National Academies identified four key considerations for maintaining U.S. competitiveness through transformational technological advances. These considerations extend beyond quantum computing and synthetic biology, and more broadly address the role of federal and nonfederal entities in supporting research for such advances. The key considerations experts identified were (1) developing a strategic approach for transformational technology, (2) fostering information sharing, (3) focusing on technology development and commercialization, and (4) strengthening the science and technology workforce.[89]

---

[89] The evidence presented in this section is primarily based on GAO's analysis of a transcript of this meeting. During the meeting, experts provided presentations, which were followed by open discussion among all meeting participants. After the end of each session, an expert summarized what was conveyed and other experts were afforded an opportunity to agree or disagree with the characterization of what was discussed. In addition, we conveyed to experts our interpretation of their statements for their input and correction. For purposes of quantifying expert remarks, we refer to a statement from an individual expert as being from one expert, and unless there is significant disagreement in the transcript of the meeting, we

Developing a Strategic Approach for Transformational Technology

Experts emphasized the importance of developing a strategic approach for advancing potentially transformational technologies for maintaining U.S. competitiveness.

Experts explained that there are multiple methods for strategic approaches and one of these experts explained that different methods can be effective for different research areas. However, experts described two key aspects of a strategic approach: (1) mechanisms to bring together potential partners such as through partnerships or consortia and (2) future-oriented technology planning such as through roadmaps or grand challenges. We further describe these aspects below:

- **Partnerships and consortia.** Experts described how partnerships and consortia provide a means of bringing together stakeholders from across different sectors—including the public, private, or academic sectors—to support development of transformational technologies. According to experts, partnerships and consortia can provide a means of strategically organizing the shared resources of different stakeholders and identifying what is needed, such as pre-competitive research to develop a technology, and how that technology might be used.[90] Experts noted examples of partnerships and consortia to support technology development:

---

refer to statements from two or more experts as being from experts. In cases of significant disagreement in the transcript, we refer to statements from two to three experts as being from a few experts, and statements from four to six experts as being from some experts. We did not ask experts to come to a consensus or agreement on the topics discussed. This section also includes information from interviews with stakeholders and agency officials and other written documentation.

[90] Pre-competitive research represents research and development activities up to the stage at which technical uncertainties are sufficiently reduced to permit preliminary assessment of a technology's commercial potential. This stage of research occurs prior to the development of application-specific technology prototypes.

Lessons Learned from the SEMATECH Consortium

In the face of international competition during the 1980s, the U.S. semiconductor industry took steps to strengthen its domestic capabilities. In one such step, U.S. semiconductor companies formed the Semiconductor Manufacturing Technology (SEMATECH) consortium in 1987 to further semiconductor manufacturing technology. SEMATECH received federal financial assistance through 1996, after which its leadership chose not to solicit continued federal support because of improvement in the U.S. semiconductor industry's competitiveness. In a September 1992 report, GAO provided lessons learned from the federal government's participation in SEMATECH and identified several considerations for future consortia, including that they:

- are industry-led and industry provides at least half of the annual funding because industry can best design a research program to meet its needs;
- develop a comprehensive industry assessment and prepare an operating plan that identifies realistic objectives and milestones as a basis for receiving federal funds;
- include active participation by member companies' senior executives in establishing research priorities and overseeing technological progress;
- have a program to improve long-term working relationships between manufacturers and key suppliers, unless inappropriate for the industry's structure;
- emphasize research projects that improve an industry's overall efficiency and that have industrywide applications;
- consider ways to provide access for smaller industry members that might not have the resources to participate; and
- establish criteria for determining how or when government should end its funding.

Sources: GAO and National Research Council of the National Academies. | GAO-18-656.

- **Manufacturing USA institutes.** One expert described the 14 Manufacturing USA institutes as a model for coordination and as the largest recent U.S. technology development project.[91] As we reported in April 2017, each institute is a public-private partnership between a sponsoring federal agency (DOD, DOE, or Commerce) and a nonfederal entity (generally a nonprofit organization or university) in charge of the institute's operations.[92] Each institute has a technological focus, such as additive manufacturing, advanced flexible electronics, or regenerative medicine, and includes members such as companies, nonprofit organizations, academic institutions, and federal agencies.
- **Semiconductor Manufacturing Technology consortium (SEMATECH).** Experts described SEMATECH, a nonprofit consortium that supported research and development on advanced semiconductor manufacturing, as a successful, industry-led, public-private collaboration that helped government and industry stakeholders take a strategic approach to challenges facing the U.S. semiconductor industry in the late 1980s.[93] However, Commerce's

---

[91] The Manufacturing USA institutes were established after a series of reports by the President's Council of Advisors on Science and Technology. The reports' recommendations included that the federal government establish a national network of manufacturing innovation institutes as public-private partnerships to create a manufacturing research infrastructure to help bridge the gap between research and development activities and domestic production. See, for example, President's Council of Advisors on Science and Technology, *Report to the President on Ensuring American Leadership in Advanced Manufacturing*, and President's Council of Advisors on Science and Technology, *Report to the President on Capturing Domestic Competitive Advantage in Advanced Manufacturing*. In 2012, at the request of the President and using existing statutory authorities, DOD, with co-funding and participation from DOE and other agencies, established a pilot manufacturing innovation institute focused on additive manufacturing. Subsequently, DOD and DOE established additional institutes using their existing statutory authorities. The Revitalize American Manufacturing and Innovation Act of 2014 called for the Secretary of Commerce to establish a Network for Manufacturing Innovation program. Pub. L. No. 113-235, § 703(2), 128 Stat. 2220, 2221 (2014) (codified at 15 U.S.C. § 278s). In September 2016, the Secretary of Commerce announced a new public name for the program, Manufacturing USA.

[92] See GAO-17-320.

[93] In prior reports, we and the National Research Council found that SEMATECH helped improve the U.S. semiconductor industry's competitive position. See GAO, *Federal Research: Lessons Learned from SEMATECH*, GAO/RCED-92-283 (Washington, D.C.: Sept. 28, 1992) and Charles W. Wessner, Editor, Board on Science, Technology, and

NIST officials noted that after federal support ended, SEMATECH began accepting memberships from companies from competitor countries, which led to a transfer of technology through the consortium's work outside the United States.[94]

- **Grand challenges, strategies, and roadmaps.** Experts described the importance of grand challenges, strategies, and roadmaps in supporting a strategic approach to developing transformational technologies.[95] In particular, experts described how these mechanisms help stakeholders coalesce around technology goals and organize efforts toward reaching them. Examples experts noted included the following[96]:
- **Brain Research through Advancing Innovative Neurotechnologies (BRAIN) Initiative.** Experts described the BRAIN Initiative, which was launched in 2013 to build neuroscience measurement tools, as a key example of a grand challenge. The BRAIN Initiative—led by HHS (specifically NIH), NSF, and DARPA, with the participation of other federal agencies as well as foundations,[97] universities, and industry—seeks to deepen understanding of the human mind and to improve how brain disorders are treated, prevented, and cured.[98]

Economic Policy, Policy and Global Affairs, National Research Council of the National Academies, *Government-Industry Partnerships for the Development of New Technologies* (Washington, D.C.: National Academies Press, 2002).

94 NIST officials cautioned that the Manufacturing USA Institutes could face similar challenges after the initial period of federal support for the institutes ends.

95 According to a 2015 report from the National Economic Council and OSTP, grand challenges are ambitious but achievable goals that harness science, technology, and innovation to solve important problems. See National Economic Council and Office of Science and Technology Policy, *A Strategy for American Innovation* (Washington, D.C.: October 2015).

96 In addition to the examples provided, other initiatives, such as the Human Genome Initiative and NIH's Precision Medicine Initiative, among others, may also provide illustrative examples.

97 One expert noted that foundations played a key role in promoting, convening, and creating consensus within the BRAIN Initiative.

98 The BRAIN Initiative is working to develop new technologies to explore how the brain's cells and circuits interact at the speed of thought, ultimately uncovering complex links between brain function and behavior, through the following goals: (1) accelerating the development and application of new neurotechnologies, (2) enabling researchers to produce a dynamic picture of the functioning brain, (3) exploring brain functionality, (4) linking brain function and behavior, and (5) advancing consumer applications.

- **National Nanotechnology Initiative.** Experts described the National Nanotechnology Initiative as a key example of a federal government strategic effort. The National Nanotechnology Initiative began in 2000 and is an interagency effort to bring together the nanotechnology-related activities of 28 federal agencies in an effort to enhance understanding and control of nanoscale material. The National Nanotechnology Initiative maintains a strategic plan describing the initiative's vision and goals and the strategies to achieve these goals. In discussing this initiative, experts described how it could enable federal agencies to share information on their research and ensure that key research areas are advanced in pursuit of a long-term national nanotechnology strategy.

Grand challenges may be articulated through strategy documents and, according to experts, involve getting stakeholders to think about potentially transformational technologies in a future-oriented way. Roadmaps, according to experts, represent detailed plans to guide progress toward a technology goal. Federal agencies, industry, or others may lead roadmapping efforts, according to experts. Additionally, one expert stated that roadmaps can help accelerate technology development. Another expert noted that for some fields, such as quantum computing and synthetic biology, a technology development strategy is needed in addition to a research and development strategy because the former outlines how a technology would move forward beyond the research and development phase.

Across both of these aspects of a strategic approach, experts emphasized the importance of a sustained commitment of resources to support technology development. One expert also emphasized the importance of setting tough performance objectives without specifying how innovators will solve a problem. Experts acknowledged that developing shared national strategies is challenging in the United States, in part because of the decentralized nature of research support across multiple federal agencies.[99] However, experts also cited as strengths of the federal research system the ability of federal agencies to support multiple approaches to developing transformational technologies in accordance with their missions and the ability to evolve and try new approaches.[100]

Experts identified several indicators of when developing a strategic approach might be important to support U.S. competitiveness through transformational technological advances in a particular area. Specifically:

- **Convergence of advances across different technology areas.** Experts described how transformational technologies often occur as a result of different technologies that have advanced incrementally over time.[101] One expert noted the development of the Global Positioning System as an example of a technology that required the convergence of advances in computing power, satellite technology, geospatial imaging, and timekeeping.[102] Because of the strength and role of the federal government in convening and fostering engagement among non-traditional collaborators on interdisciplinary issues, experts identified technology convergence as a potential indicator of the need to take a strategic approach.
- **Progress from discovery to real-world application.** Experts described how progress from discovery in an area of science to the appearance of niche applications for a technology can be an indicator of the need to take a strategic approach. According to one

---

99 One expert stated that establishing priorities as policy mandates—such as through OSTP-led initiatives or legislation—will cause federal agencies to prioritize these efforts and coordinate relevant activities.

100 The National Research Council also cited the decentralized nature of the federal research system as a strength in a 2012 report. See Committee on Cooperative National Innovation Policies: Best Practice for the 21st Century; Board on Science, Technology and Economic Policy; Policy and Global Affairs, National Research Council of the National Academies, *Rising to the Challenge: U.S. Innovation Policy for the Global Economy* (Washington, D.C.: National Academies Press, 2012).

101 Incremental technology improvements refer to updates or improvements to existing products and technologies.

102 According to one expert, cross-agency mechanisms that enable cross-disciplinary research and development—which is often called convergence—are an important way to bring together a mix of relevant scientific fields.

expert, one challenge in technology development is how to push the technology forward as quickly as possible to develop it into something useful. Experts explained that by taking a strategic approach that extends beyond early-stage research, the federal government can support the development of potentially transformational technologies.

- **Existence of barriers to technology development.** Experts identified several barriers to the development of transformational technologies that could indicate the need to take a strategic approach to developing a technology. Examples of barriers experts identified included high capital costs for research, prototyping, demonstration, or other aspects of a technology development life cycle; regulatory barriers; lack of consensus on standards; and technology measurement challenges, such as limitations in the availability of tools with which to measure products or processes.[103] Experts described multiple ways in which the federal government can play an important role in addressing such barriers through helping efforts to de-risk technologies, establishing or revising regulations, supporting standards development, and developing measurement tools.[104]
- **Increasing involvement across multiple stakeholders or competitors.** Experts described aspects of how increasing involvement across multiple stakeholders in a particular technology area can indicate the need to take a strategic approach to developing a transformational technology. For example, when multiple federal agencies are working in a technology area or industrial participants increase involvement in a particular

---

[103] Experts described how regulations could present barriers to the development of potential transformational technologies in different ways. For example, some experts described how regulations, particularly when not updated, may hinder technology innovation. Experts also described how gaps in regulatory science—which has been defined as the development and use of new tools, standards, and approaches to more efficiently develop products and effectively evaluate product safety, efficacy, and quality— may present a barrier.

[104] However, one expert cautioned that standards should not be developed too early because they can deter the growth of alternative technology pathways or lock development into existing pathways.

technology, experts said such involvement could signal that a strategic approach is needed to work across boundaries and engage the research community in a coordinated way. Similarly, according to one expert, increasing international competition in a technology area could serve as an indicator of the need for the federal government to exercise leadership through a strategic approach to organize domestic public and private efforts in order for the United States to remain competitive.

- **Need for sustained, long-term investment in areas of national interest.** Experts identified the need for sustained, long-term investment in areas of national interest as a potential indicator of the need for a strategic approach to transformational technologies. Experts described how the short-term cycles of many federal programs and disincentives for the private sector to sustain long-term investments can present challenges to developing transformational technologies, which one expert noted can take years or even decades to develop. Experts also cited a need for a strategic approach to advancing a technology when it has the potential to be transformational and presents enormous societal benefits.

In the areas of quantum computing and synthetic biology, experts cited a need to develop a strategic approach to maintain U.S. competitiveness. Within the area of quantum computing, experts cited all of the indicators identified above in stating that U.S. competiveness in quantum computing could benefit from a national strategy. For example, experts described the need to foster interdisciplinary engagement across the fields of physics, engineering, and computer science to support convergence of advances in these areas to further quantum computing technology. Experts also indicated that real-world applications are beginning to become apparent in the area of quantum computing. However, they noted that significant barriers to development exist and discussed a need for sustained longterm investment in this area, which has significant implications for national security, and according to one expert, economic competitiveness.

Moreover, experts expressed concern over the significant and increasing international competition from China, the European Union, and other countries. One expert noted that given the security implications of quantum computing technology, the United States needs to find a way to counter the significant investment that China is making. Stakeholders and one agency official we interviewed cited similar concerns, such as the European Union's plans to launch a flagship initiative on quantum technology, which includes quantum computing; therefore, the United States needs a national quantum computing strategy, the experts said.

Similarly, with regard to synthetic biology, experts cited several of the indicators described above in stating that the United States could benefit from a strategic approach to maintain competitiveness. For example, experts discussed barriers to technology development, including a lack of measurement tools and regulatory barriers. According to one expert, before the 2017 update to the Coordinated Framework for the Regulation of Biotechnology, the system was last updated in 1992.[105] The expert said that it was not yet clear if the updated framework would help advance synthetic biology research. Experts also noted the need to engage across multiple stakeholders in this area; in particular, one expert noted the need for leadership to advance a dialogue about how synthetic biology could help address issues of national concern.[106] Experts described significant foreign

---

[105] The Coordinated Framework for the Regulation of Biotechnology, a policy originally announced in 1986, outlined a comprehensive federal regulatory policy for ensuring the safety of biotechnology research and products. 51 Fed. Reg. 23,302 (June 26, 1986). In 2015, the Executive Office of the President issued a memorandum directing the primary agencies that regulate the products of biotechnology—EPA, the U.S. Food and Drug Administration, and USDA—to update the Coordinated Framework for the Regulation of Biotechnology by clarifying roles and responsibilities, to develop a long-term strategy to ensure that the federal biotechnology regulatory system is prepared for future biotechnology products, and to commission an expert analysis of the future landscape of biotechnology products to support these efforts. The agencies issued the updated framework in 2017.

[106] As part of a November 2016 workshop on Making the Living World Engineerable: Science, Practice, and Policy, participants discussed current trends in synthetic biology and implications for federal policy. Participants described similar challenges in terms of the barriers to advancing the technology and the need to increase engagement across multiple federal agencies and stakeholders in a strategic approach to how synthetic biology could be developed to address U.S. needs. Forum on Synthetic Biology; Committee on Science, Technology, and Law; Policy and Global Affairs; National Academies of Sciences, Engineering, and Medicine, *Making the Living World Engineerable: Science, Practice, and*

competition in synthetic biology. One expert said that there are more than 40 countries that have a unified strategy for synthetic biology.[107] While one expert stated that NSF has initiated a synthetic biology roadmapping effort, a few experts stated that the United States does not have a similar unified synthetic biology strategy.[108] One expert said that in the absence of such a strategy, the United States faces economic and physical security risks. Stakeholders we interviewed raised similar concerns.

Fostering Information Sharing

Experts also suggested considering how to foster information sharing to help maintain U.S. competitiveness through transformational technological advances. Experts discussed the role the federal government can play in bringing together stakeholders to discuss emerging technologies and collaborate on pre-competitive research. For example, according to one expert, in 2015, 2 years after the BRAIN initiative was launched, the White House convened a meeting that brought together industry partners, academic researchers, and government scientists to share information and discuss research plans.[109] This expert highlighted the importance of communication among representatives of organizations that would not normally work together, and how these conversations about where they saw research going over the next 5 years led to greater understanding and collaboration to support the research under this initiative.

Policy: Proceedings of a Workshop in Brief (Washington, D.C.: National Academies Press, December 2016).

[107] Many countries have developed strategies to commercialize and industrialize technological advances, according to a 2012 National Research Council report. See National Research Council of the National Academies, *Rising to the Challenge*.

[108] This is consistent with the findings of a 2012 report by the National Research Council, see National Research Council of the National Academies, *Rising to the Challenge*.

[109] We have previously reported that in-person interactions help to build trust—an essential element to collaborative relationships—and strengthen professional networks. See, GAO, *Managing for Results: Implementation Approaches Used to Enhance Collaboration in Interagency Groups*, GAO-14-220 (Washington, D.C.: Feb. 14, 2014).

Experts identified three key reasons for sharing information to facilitate transformational technological advances in supporting U.S. competitiveness:

- **Convergence of different disciplines.** Experts generally agreed that information sharing can facilitate an interdisciplinary approach to study a problem, which they said is important to the nation's ability to conduct research for transformational technological advances. The federal government's ability to convene groups, according to one expert, is particularly important for interdisciplinary areas of study because it can help bring stakeholders together to discuss how research could help address an area of national need. Another expert explained that agencies' research is increasingly interdisciplinary, which increases the importance of coordinating across agencies.[110] Agency officials and stakeholders we interviewed also discussed the importance of sharing information across fields of study. One stakeholder said that without government funding for interdisciplinary efforts in quantum computing, it will be challenging to solve problems, such as creating some of the computer programming needed to operate a quantum computer, that need to be solved in order to make quantum computing viable.
- **Overcoming barriers to innovation.** Experts discussed how information sharing can facilitate the identification of barriers to innovation and help overcome them. For example, one expert noted the importance of information sharing in trying to address the challenges the U.S. semiconductor industry faced in the 1980s. The expert emphasized the recognition that individual companies

[110] This view is consistent with a National Research Council study that stated that the scientific opportunities facilitated by the convergence or coming together of insights and approaches from originally distinct fields of study can make fundamental contributions to solve society's most difficult problems. See Committee on Key Challenge Areas for Convergence and Health; Board on Life Sciences; Division on Earth and Life Studies; National Research Council of the National Academies, *Convergence: Facilitating Transdisciplinary Integration of Life Sciences, Physical Sciences, Engineering, and Beyond* (Washington, D.C.: 2014).

could not address the barriers to innovation on their own and that they needed information sharing, such as cross-licensing of intellectual property and communication about roadmapping to overcome barriers that they faced. Another expert explained that information sharing across federal agencies led to the identification of the U.S. biotechnology regulatory system as a significant barrier to innovation and that, based on this, the Coordinated Framework for the Regulation of Biotechnology was updated.[111] This expert further said that information sharing is the first step in coordination—by sharing information, agencies can determine where there might be overlapping research efforts or gaps in ongoing research.

- **Leveraging international research.** Experts explained that bringing technologies to the United States that were developed elsewhere is not something that has been central to U.S. science and technology policy, but they stressed that the United States needs to consider how to take advantage of research that other countries are conducting and effectively utilize that information to maintain U.S. competitiveness. For example, one expert described the importance of the iGEM competition as an opportunity for information exchange among researchers from around the world who are working in synthetic biology-related fields. In describing this example, the expert noted that most bioengineers will not be U.S.-based and that, to remain competitive in synthetic biology, the United States needs to better understand discoveries being made by researchers from around the world.

---

[111] As noted previously, the 2015 Executive Office of the President memorandum that led to the 2017 Update to the Coordinated Framework for the Regulation of Biotechnology stated that the update should clarify roles and responsibilities of the primary agencies involved in regulating biotechnology products. Specifically, the memorandum provided that the update should, among other things, clarify the mechanism and timeline for regularly reviewing, and updating as appropriate, the Coordinated Framework to minimize delays, support innovation, protect health and the environment, and promote public trust in the regulatory systems for biotechnology products.

Experts said that while information sharing is important, there are tradeoffs, particularly with regard to sharing and protecting pre-competitive intellectual property. The experts said that the benefits of sharing pre-competitive intellectual property include the opportunity to speed innovation by allowing multiple researchers to work with the intellectual property concurrently and by preventing foreign competitors from restricting use of the intellectual property through obtaining a patent. Economically valuable knowledge can spread through publicly and freely available records such as scientific publications and open source software. Such knowledge can be used repeatedly, can quickly spread to users outside the institutions where it was created, and can lead to the creation of new products. For example, one expert stated that, as of October 2017, a quantum computer we described earlier in this report had been available over the Internet for public use for about a year and had 50,000 users. Having a larger number of users working with this resource could lead to more rapid discovery of ways in which a quantum computer might be used than if it had not been shared. The expert said that because this technology exists, it should be developed as quickly as possible to determine what its first useful application will be and to find the first problem that only a quantum computer can solve. Doing so, the expert said, would create opportunities in which a U.S. company could profit from the technology while also developing it. In addition, information sharing was cited as instrumental to the success of the Human Genome Project, according to NIH officials we interviewed, because the project made the genome's sequencing available as a resource for researchers to use.[112]

Human Genome Project

[112] The Human Genome Project, coordinated by DOE and NIH, was an international, collaborative research project with a goal of completely mapping and understanding all the genes of human beings, collectively known as the genome. The project completed and published the full human genome sequence in 2003.

A deoxyribonucleic acid (DNA) strand around the outline of a person.

The Human Genome Project, which formally began in 1990, was a 13-year international collaborative research project coordinated by the Department of Energy and the National Institutes of Health. The Human Genome Project's goals were to (1) identify all the genes in human DNA, (2) determine the chemical base pair sequences of human DNA, (3) store this information in databases, (4) improve data analysis tools, (5) conduct technology transfer, and (6) address the ethical, legal, and social issues that may arise from the project. The full sequence of the human genome was completed and published in April 2003.

Through its policy of open data release, the Human Genome Project facilitated the research of others. The Human Genome Project also anticipated and promoted commercializing genomic resources and applications by establishing an infrastructure and supporting private-sector technology development. Consequently, the project led to new tools to support biological research. Further, the data and technologies generated by the project and related research present a broad array of commercial opportunities across many areas of the economy. These include more individualized diagnostics, prognostics, drugs, and other therapies as well as hardier, more nutritious, and healthier crops and animals, among other applications.

Sources: Department of Energy and the National Institutes of Health (text); Department of Energy (image). | GAO-18-656.

At the same time, experts said that while information sharing is important, there are risks, such as foreign commercialization of U.S.

intellectual property. Experts noted that the world is increasingly competing with the United States in research for transformational technological advances. One expert cautioned that while information sharing is important for transformational technologies, it must be done carefully so that other companies do not exploit a technology or it is not leaked to a foreign competitor. Similarly, one stakeholder said that while information sharing is beneficial at the early stages of technology development, a balanced approach to information sharing—an approach that allows for trade secrets and that guards some research results—is needed once a technology is no longer in the early stages of development.

In light of these tradeoffs, experts emphasized the importance of ensuring that intellectual property protections support U.S. competitiveness; however, they also described challenges with how intellectual property is managed in the United States. For example, experts said it can be challenging to bring industry and academic researchers into partnerships that support transformational technological advances. Experts explained that some collaborators are willing to openly share their intellectual property, while other experts noted that some collaborators may be less inclined to do so because they view intellectual property as a profitable commodity. Additionally, one expert cited differences between potential industry and academic collaborators' knowledge of, and attention paid to, developing technologies into commercial products as a potential barrier. One expert said that foreign countries generally allow university-developed intellectual property to be owned and licensed by the inventors or third-party companies (instead of the university). This can create a foundation for a startup company or make it easier to get the interest of companies who would like to acquire a university-based technology or process. The expert noted that in one circumstance, this has given an advantage to a foreign university in recruiting top researchers, helping it to become a leader in quantum computing. However, another expert stated that most major research universities have moved to a model of developing partnerships with firms, especially startups, which has minimal upfront licensing costs, and shared gains over time if the project is successful—

according to that expert, such universities typically share research intellectual property rights with faculty inventors.

Focusing on Technology Development and Commercialization

Focusing on technology development and commercialization is another policy consideration that experts identified for maintaining U.S. competitiveness through transformational technological advances. According to experts, the United States' "innovation ecosystem"—the network of public and private institutions within a country whose activities and interactions initiate, develop, commercialize, and diffuse new technology innovations—has either lost or needs better mechanisms for commercializing technologies to maintain U.S. competitiveness. To address this issue, experts discussed how the federal government could focus on technology development and commercialization by providing support across multiple stages of innovation and support for the development of tools to enhance innovation.

Providing Longer-Term Assistance to Support Technology Development

Experts discussed a need to improve technology development and commercialization by providing support across multiple stages of innovation. Experts described how sustained federal research investments have led to key scientific discoveries, including, for example, NIST and IARPA's decade-long support for quantum computing research and NSF's investment in synthetic biology. However, while experts said federal agencies' ability to support new discoveries is a strength, they explained that the United States is losing the ability to commercialize technologies that are invented here. For example, according to one expert, while the technology might soon be available to build small (100 qubit) quantum computers, the United States does not have the necessary enterprise in place to manufacture those systems. Experts stated that it may take decades or more from the time research is funded until it is commercialized. During this intervening period, significant investment is needed to support the innovation cycle in terms of research in the design, building, and testing of new product prototypes and production processes.

Experts described an increasing reliance, over time, on venture capital funding to support investments in the innovation cycle. They said that while this is generally working well in some areas such as software and biotechnology, venture capital investors have become less willing to support other technologies that require higher levels of capital investment, longer-term returns, and greater risk. For example, one expert stated that while the U.S. venture capital system spends $70 billion annually on technology commercialization activities, in 2015, the expert estimated that 5 percent of venture capital funding went to hard technologies.[113] Multiple reports in recent years have documented the challenges associated with how the innovation cycle is supported in the United States and its implications for the domestic commercialization and production of new technologies.[114] For example, in a 2012 report, the National Research Council stated that discoveries and inventions originating from research conducted at U.S. universities, corporations, and national laboratories no longer naturally led to products that are commercialized and manufactured within the United States. According to this report, manufacturing is important in developing new products because in many high-technology industries, design cannot easily be separated from manufacturing, and a

---

[113] According to NSF's *Science and Engineering Indicators 2018*, in the United States, venture capital early- and later-stage investment reached $65 billion between 2013 and 2016. National Science Board, *Science and Engineering Indicators 2018.* Also, a recent report by MForesight—a national consortium that engages advanced manufacturing stakeholders—showed that the majority of venture capital investment in 2017 (57.4 percent) went to software while a relatively smaller proportion (1.5 percent) went to industrial and energy technologies. See Sridhar Kota and Thomas C. Mahoney, *Manufacturing Prosperity: A Bold Strategy for National Wealth and Security,* MF-TR-2018- 0302 (Alliance for Manufacturing Foresight, June 2018). Hard technologies are physical products obtained from scientific breakthroughs that must be translated from science into commercially viable products and require time and resources to develop. Examples of hard technology breakthroughs include technologies that increase the efficiency of power systems and production processes, and sustainably produced chemicals.

[114] See, for example, Peter L. Singer and William B. Bonvillian, *innovation Orchards": Helping Tech Start-Ups Scale* (Information Technology and Innovation Foundation, March 2017); Bonvillian and Singer, *Advanced Manufacturing*; Kota and Mahoney, *Manufacturing Prosperity*; and Berger, *Making in America.*

lack of sustained investment in research and infrastructure threatens to damage the U.S. innovation ecosystem, economy, and security.[115]

To address this issue, experts discussed a need to provide longer-term federal financial assistance to better support technology development across multiple stages of innovation. Experts stated that federal agencies often support research on short-term funding cycles (e.g., 3 years or less) that may not be conducive to the long-term support sometimes needed to effectively de-risk potentially transformational technologies. A 2017 National Academies report cited short-term funding as one factor that has resulted in U.S. science losing its flexibility and nimbleness, elements that feed new discovery.[116] Additionally, experts said that federal agencies' support may not extend to the later stages of technology development but providing longer-term support for research is an important part of the federal government's role in advancing transformational technologies. For example, one expert said that long-term federal support facilitates creating a research infrastructure that can support a technology's development.

Experts cited several examples of how federal agencies' programs provide different models for supporting technology development across multiple stages of innovation.

- **Advanced Technology Program.** Experts cited NIST's Advanced Technology Program—which COMPETES 2007 repealed—as a success in terms of its efforts to support transformational research.[117] Experts cited several aspects of the program in discussing its success, including its support for (1) research that

[115] See National Research Council of the National Academies, *Rising to the Challenge.*

[116] National Academies of Sciences, Engineering, and Medicine, *Opportunities in Intense Ultrafast Lasers.*

[117] The program's purpose was to help U.S. businesses create and apply the generic technology and research results necessary to (1) commercialize significant new scientific discoveries and technologies rapidly and (2) refine manufacturing technologies. The Advanced Technology Program was established by the Omnibus Trade and Competitiveness Act of 1988, Pub. L. No. 100-418, § 5131(a), 102 Stat. 1107, 1439 (1988), and was repealed by COMPETES 2007, Pub. L. No. 110-69, § 3012(a), 121 Stat. 572, 593 (2007), which replaced the program with the Technology Innovation Program. The Technology Innovation Program was repealed in 2017.

accelerated the development of high-risk technologies with the potential for broad-based economic benefits to the nation; (2) information sharing across different sectors; (3) active project management and workshops that taught awardees how to pitch their technology to venture capital investors, according to one expert.[118] One expert noted that the program collaborated with NIH to develop diagnostic approaches that advanced the genomic revolution.

- **ARPA-E.** Experts described ARPA-E—which was modeled after DARPA—as an important challenge-based federal effort to advance technologies in areas aligned with DOE's mission.[119] Aspects of the ARPA-E model one expert cited as important to the program's ability to support transformational technological advances included, among others, support for higher-risk research and the autonomy that program directors have in seeking expert input and selecting research projects.[120]

---

[118] One expert clarified that the Advanced Technology Program did not provide awardees with financial assistance to commercialize technologies.

[119] As specified by statute, ARPA-E's program goals are to enhance the nation's economic and energy security through the development of energy technologies and to ensure that the United States maintains a technological lead in developing and deploying advanced energy technologies. 42 U.S.C. §16538(c)(1).

[120] The National Academies conducted an assessment of ARPA-E under a Congressional mandate. Specifically, COMPETES 2007 provided that after ARPA-E had been operational for 6 years, the Secretary of Energy was to offer to enter into a contract with the National Academy of Sciences for it to conduct an evaluation of how well the agency was achieving its mission and goals. 42 U.S.C. § 16538(l). The National Academies issued its report in 2017. See Pradeep K. Khosla and Paul T. Beaton, Editors, Committee on Evaluation of the Advanced Research Projects Agency-Energy; Board on Science, Technology, and Economic Policy; Policy and Global Affairs; Board on Energy and Environmental Systems; Division on Engineering and Physical Sciences; National Academies of Sciences, Engineering, and Medicine, *An Assessment of ARPA-E* (Washington, D.C.: National Academies Press, 2017). GAO has also reported on ARPAE. See GAO, *Department of Energy: New Process to Review Financial Assistance for Research Projects Created Uncertainty*, GAO-18-278 (Washington, D.C.: Feb. 28, 2018); and GAO, *Department of Energy: Advanced Research Projects Agency-Energy Could Benefit from Information on Applicants' Prior Funding*, GAO-12-112 (Washington, D.C.: Jan. 13, 2012).

- **Manufacturing USA Institutes.** One expert described the Manufacturing USA institutes as an important federal effort to support emerging technologies across multiple stages of innovation. Another expert explained that in order to continue to capture the economic benefits of the innovation system, the United States needs to embed the knowledge for technology production locally within the country. The first expert said the Manufacturing USA institutes help increase the connectivity among different actors involved with specific technology areas and improve their ability to leverage advances in those areas.

Experts also discussed how other countries' long-term funding for research efforts may help them support technology development. For example, one expert discussed Germany's Fraunhofer Institutes, where the government makes research investments over time frames of 5 or even 20 years and rewards successful projects with funding increases each year.[121]

In addition, one expert noted that other countries such as the Netherlands and Singapore also provide long-term research funding, allowing them to develop the broader research infrastructure necessary to support technology development. In the area of quantum computing, one expert stated that the Netherlands' investment has contributed to one of the largest quantum computing-focused efforts in the world. According to one expert, if U.S. researchers do not conduct the research necessary over the long term to prove their research ideas, other countries will have the opportunity to pick up where U.S. researchers leave off and commercialize technologies based on this research.

---

[121] According to Fraunhofer's 2016 annual report, they conduct applied research that is intended to drive economic development and benefit society. Projects are initially established for 5 years and, at the end of this period, the projects' performance is evaluated to determine eligibility for additional funding. Fraunhofer-Gesellschaft, *Annual Report 2016 Embracing Digitalization*, (Munich, Germany: 2017). One expert described the Fraunhofer Institutes as using a shared technology development model between industry, engineering schools, and federal and state governments.

Supporting Development of Tools to Enhance Innovation

3D Bioprinting

A bioprinted coronary artery.

3D bioprinting is a tool that scientists are developing in the field of regenerative medicine. 3D bioprinting uses 3D printing with biological materials to create skin, bones, arteries, and a variety of other tissues and organs. For example, the Department of Defense has conducted research into using 3D bioprinting to repair skin damaged by burns—injuries that account for 10 to 30 percent of battlefield casualties. To repair burned skin, researchers have created scans of burns that a computer then uses to have a 3D printer reconstruct the burned skin.

3D bioprinting has also been used to create small blood vessel networks that contain living cells that have joined with the blood vessel networks in a mouse, allowing blood to circulate through them. Such printed blood vessels could be used to replace a damaged heart muscle. In the future, such organs could be grown using 3D bioprinting and the cells of the person who needs the organ, and they could be used in place of transplanted organs. 3D bioprinted tissues could also be used to test the safety of new drugs. 3D bioprinting is in the early stages of development.

Sources: Department of Defense and National Institutes of Health (text); Adam Feinberg, Carnegie Mellon University (image). | GAO-18-656.

Experts stated that tool development is critical to transformational technological advances and discussed a need for federal government support for tool development to maintain U.S. competitiveness. A tool is something—such as equipment used for a specific purpose, a modified biological system, or a computer program—that is used to perform a task or that is needed to practice a profession. According to one expert, tools are crucial supporting technologies that are necessary for the product development process. According to recent reports, research in tools development can lead to the introduction of new products, materials, or the ability to produce materials at the commercial level.[122]

Experts explained that the United States is at risk of losing its ability to develop tools, and they identified challenges to tool development, including the following:

- **Unclear needs and long time frames.** According to experts, industry may be less likely to invest in tool development when tools do not support existing products, but, rather, are a part of solving technology challenges that are not clearly defined. In this context, experts explained that tool development can take a relatively long time, which may not be compatible with industry's short innovation time frames.
- **Potentially high or unrecoverable costs.** Developing tools is expensive, according to experts, and when creating a new tool, companies have to consider whether they will be able to recover their costs. One expert described a circumstance in which a modified laser was needed to support research on a quantum system. The expert explained that a laser manufacturing company would need to change its production line in order to make the

[122] See National Research Council, *An Assessment of the National Institute of Standards and Technology Material Measurement Laboratory: Fiscal Year 2014* (Washington, D.C.: 2015), and Subcommittee for Advanced Manufacturing of the National Science and Technology Council, *Advanced Manufacturing: A Snapshot of Priority Technology Areas Across the Federal Government* (Washington, D.C.: April 2016).

modified laser, and it would be very expensive for the company to adjust its production line to make only the modified laser.

Experts emphasized the important role federal agencies can play in helping overcome these challenges to tool development. For example, experts described the importance of federal support for developing measurement tools to accelerate and improve the learning cycles around designing, building, and testing technologies and products. Experts specifically cited NIST's role in the development of measurement tools. For example, through the NIST-on-a-Chip program NIST is developing ultra-compact, inexpensive tools that will measure quantities such as time, distance, current and voltage, and temperature and pressure and that will allow measurement technologies to be deployed without requiring traditional measurement services. In line with NIST's goals, the private sector will manufacture and distribute these technologies.[123] Experts also noted the important role federal agencies play in providing access to tools, such as technology testbed facilities to support de-risking technologies through prototyping and other development activities.

STRENGTHENING THE SCIENCE AND TECHNOLOGY WORKFORCE

Experts identified strengthening the science and technology workforce as a consideration for maintaining U.S. competitiveness through transformational technological advances. According to experts, there is a need for federal agencies to work with academia and industry to improve connections between the training academia provides and what industry needs, such as interdisciplinary training. Experts further discussed the recruitment of researchers and the retention of research talent and a

---

[123] We recently reported on the measurement services NIST provides to support U.S. industry's needs, among other things. See GAO, *National Institute of Standards and Technology: Additional Review and Coordination Could Help Meet Measurement Services Needs and Strengthen Standards Activities*, GAO-18-445 (Washington, D.C.: July 26, 2018).

technically trained workforce; according to experts, attracting researchers has historically been a U.S. strength, but this ability may be at risk.

Improving Connections between Academic Training and Industry Needs

Experts identified the need to improve connections between academic institutions and industry so that the training academia provides corresponds to industry's needs, particularly for interdisciplinary research fields. Without strengthening these connections, according to experts, academia may not deliver the interdisciplinary training needed for some research areas. Experts identified the systems engineering training needed to build a quantum computer as one such area of interdisciplinary training.[124] For example, one expert said engineers are usually unfamiliar with the quantum mechanics used in a quantum computer and this is challenging since knowledge of both disciplines—quantum mechanics and engineering—is necessary to develop the technology. Also, not many quantum computing researchers are trained in the fields of computer science or engineering, according to stakeholders and agency officials we interviewed. A few experts said that because universities are not training the researchers needed in some interdisciplinary areas, there are not enough researchers in those areas available for industry to hire.

Experts, other stakeholders and agency officials we interviewed, as well as some recent reports, identified several factors that may contribute to a disconnect between academic training and industry needs. For example, experts explained that universities appear to operate on the assumption that industry, not universities, must teach students the practical skills needed to be productive members of an engineering team.[125]

---

[124] According to the NASA Systems Engineering Handbook, systems engineering is a multidisciplinary approach for the design, realization, technical management, operations, and retirement of a system that includes the hardware, software, facilities, and personnel, among other things, required to produce results. It is a holistic discipline, in which multiple fields' contributions are evaluated and balanced to produce a system that is not dominated by a single field's perspective. It is a way to look at the big picture. *NASA Systems Engineering Handbook,* NASA SP-2016-6105 Rev 2 (Washington, D.C.: 2016).

[125] Participants expressed similar observations in a 2016 National Academies workshop on the U.S. science, technology and engineering workforce, which stated that there is often a significant gap between the knowledge, skills, and abilities most often sought by employers, and those that students bring to the workforce upon graduation. See Joe Alper, Rapporteur;

Additionally, according to a 2012 report by the National Research Council, job markets and careers for doctoral scientists and engineers have shifted since 1990 so that more than 50 percent of new doctorates work outside of academia, but there are few incentives to motivate graduate programs to align doctoral education with evolving employment activities.[126] According to one expert, graduate education is largely supported by federally funded research awards to universities which tend to support basic research, not applied research or development. This expert further stated that as a result, graduate students are not taught later stage applied work relevant to industry because that has not been what federal research has historically funded.[127] According to a different 2012 National Research Council report, cultural barriers often separate industry from academia and are reinforced by organizational incentives— universities have traditionally emphasized the need to publish research, not commercialize it.[128] Further, one expert, a stakeholder, and an agency official we interviewed said that universities generally were not hiring faculty who focus on quantum computing as part of their computer science and engineering departments. The expert attributed this to limited funding available to support those research programs. According to this expert, the financial assistance federal research programs provide can send an important signal to universities that can lead to evolving academic programs and hiring in interdisciplinary fields. A 2016 MIT report made similar observations and said that many universities remain siloed along departmental lines and need resources and structures that allow for team teaching—two people from different research areas co-teaching a course—or research in which students from

---

Board on Higher Education and Workforce; Policy and Global Affairs Division; National Academies of Sciences, Engineering, and Medicine, *Developing a National STEM Workforce Strategy: A Workshop Summary* (Washington, D.C.: 2016).

[126] The report indicated that, while the nation's scientific workforce needs had evolved over the last several decades with changes in the work of science-based industries, government agencies, and nonprofits, most research universities had not yet adequately adapted to the new realities of these labor markets. Committee on Research Universities, Board on Higher Education and Workforce, Policy and Global Affairs, National Research Council, *Research Universities and the Future of America: Ten Breakthrough Actions Vital to Our Nation's Prosperity and Security* (Washington, D.C.: 2012).

[127] According to this expert, the advanced manufacturing institutes could constitute a departure from the historical federal research model of typically funding basic research at universities.

[128] National Research Council of the National Academies, *Rising to the Challenge.*

different disciplines could be paired to answer a research question.[129] However, in synthetic biology, one expert noted that some universities have started entirely new Departments of Bioengineering because aspects of synthetic biology contribute to the development of an independent, distinctive, and complementary type of engineering. This has resulted in the development of a new curriculum that incorporates synthetic biology into the training and development of bioengineers, according to this expert.

Recruitment and Retention of Talent

Experts discussed the importance of recruiting researchers and retaining talent and a technically trained workforce. Experts stated that attracting researchers to come and stay in the United States has historically been a national strength. The Congressional Budget Office has reported that foreign-born workers contribute disproportionately to innovation.[130] Further, according to this report, foreign-born researchers account for a disproportionate number of the scientific researchers who yield many of the big discoveries and conceptual breakthroughs that drive science.[131]

However, according to a few experts, and a National Research Council report, the United States is increasingly competing with other countries to recruit and retain talented researchers.[132] Countries such as Canada, China, and Singapore are attracting talented researchers to their universities and research institutes by offering high salaries and the opportunity to run well-funded programs, according to a National Research Council report.[133] For example, according to a few experts, China started the Thousand Talents

---

[129] Phillip Sharp, Susan Hockfield, and Tyler Jacks, *Convergence: The Future of Health* (Cambridge, MA: June 2016).

[130] Congressional Budget Office, *Federal Policies and Innovation* (2014).

[131] In 2015, according to the National Science Board, 13 percent of the population of the United States was foreign born, but this portion of the population accounted for 29 to 30 percent of college-educated workers employed in science and engineering occupations and 42 to 45 percent of the workers with doctorates. National Science Board, *Science and Engineering Indicators 2018.*

[132] National Research Council of the National Academies, *Research Universities and the Future of America: Ten Breakthrough Actions Vital to Our Nation's Prosperity and Security* (Washington, D.C.: 2012).

[133] National Research Council of the National Academies, *Rising to the Challenge.*

Program in 2008 to get talented researchers to return to China.[134] The Thousand Talents Program's goal is to bring top talent trained overseas to China on a full- or part-time basis. One expert gave the example of a university president resigning from a U.S. university because he believed the possibilities for research were greater in Asia. According to one expert, the nation's ability to recruit and retain researchers may be at risk because the United States is not working to retain and incentivize talent. According to that expert, this puts the nation at risk of missing out on the next global transformational technological advance.

According to some experts, one challenge to retaining talent in the United States is that limited job opportunities are available to young researchers trained in certain areas. It is important to create conditions for young researchers to find employment in research and development, according to one expert, so that they can contribute to these areas. Creating the right incentive structure for people to produce transformational technologies in the United States is important, according to another expert, because when technologies are produced in the United States, the skills needed to produce them become embedded in that community. We have previously reported that too much location of skilled manufacturing jobs abroad can, in general, put the United States at a disadvantage in terms of its ability to design new products, according to participants in a 2013 forum on nanomanufacturing.[135] Similarly, in a 2012 report, the National Research Council stated that manufacturing is integral to new product development, and production lines are linked to an iterative innovation chain that includes research and development, product refinement, and full-scale production.[136] In many high-technology industries, design cannot

---

[134] According to a research report prepared on behalf of the U.S.-China Economic and Security Review Commission, the Thousand Talents Program is administered by various Chinese entities, including local governments. According to the report, the program administered by the University of the Chinese Academy of Science requires that candidates hold a professorship at a distinguished international university or an equivalent international research institution. See Richard P. Suttmeier, *Trends in U.S.-China Science and Technology Cooperation: Collaborative Knowledge Production for the Twenty-First Century?* Research Report Prepared on Behalf of the U.S.-China Economic and Security Review Commission, September 11, 2014.

[135] GAO-14-181SP.

[136] National Research Council of the National Academies, *Rising to the Challenge.*

be easily separated from manufacturing, and talent availability is the most important factor for deciding where to place a production facility. In some cases, according to this 2012 report, companies are choosing to produce abroad because of concerns related to the capacity of the U. S. supply chain, technical skills of U.S. workers, and the investment climate for high-volume manufacturing. Also according to this report, as a result of these factors, the United States is finding it increasingly difficult to capture the economic value generated by public and private investments in research and development.

CONCLUSION

Federal support for research in areas such as quantum computing and synthetic biology can help promote U.S. competitiveness in the global economy. For example, advances in quantum computing have the potential to lead to transformational advances in national security technologies or technology areas that rely heavily on simulation, such as pharmaceuticals and materials science for advanced manufacturing. Research in synthetic biology could help achieve significant advances in health care, energy, and other sectors. When agencies collaborate on their research efforts, they can produce more public value than when they act alone. Moreover, collaboration through mechanisms such as interagency groups can help address complex issues, such as those remaining to be resolved in quantum computing and synthetic biology. Collaboration can also mitigate challenges associated with fragmentation of efforts across multiple agencies, as well as potential overlap and duplication.

NSTC and federal agencies have taken steps, building on earlier efforts, to coordinate their activities in the areas of quantum computing and synthetic biology. Specifically, both the new QIS Subcommittee and the new synthetic biology working group have taken initial steps to implement certain leading practices that can enhance and sustain collaborative efforts. For example, both have taken steps toward agreeing on roles and responsibilities. These steps could help address problems identified in previous interagency coordination efforts. However, both the subcommittee and working group are recently established and have had limited time to fully implement the leading practices that we describe in this report. As the subcommittee and working group move forward, by taking steps to fully implement these leading practices for collaboration, member agencies could better marshal their collective efforts to support research in quantum computing and synthetic biology and help maintain U.S. competitiveness through transformational technological advances.

RECOMMENDATIONS FOR EXECUTIVE ACTION

We are making a total of five recommendations, including one to OSTP, one to Commerce, one to DOE, and two to NSF.

- As the QIS Subcommittee moves forward, the Office of Science and Technology Policy co-chair, in coordination with other co-chairs and participating agency officials, should take steps to fully implement leading practices that enhance and sustain collaboration. (Recommendation 1)
- As the QIS Subcommittee moves forward, the Department of Commerce co-chair, in coordination with other co-chairs and participating agency officials, should take steps to fully implement leading practices that enhance and sustain collaboration. (Recommendation 2)
- As the QIS Subcommittee moves forward, the Department of Energy co-chair, in coordination with other co-chairs and participating agency officials, should take steps to fully implement leading practices that enhance and sustain collaboration. (Recommendation 3)
- As the QIS Subcommittee moves forward, the National Science Foundation co-chair, in coordination with other co-chairs and participating agency officials, should take steps to fully implement

leading practices that enhance and sustain collaboration. (Recommendation 4)

- As the Interagency Working Group on Synthetic Biology moves forward, the Director of the National Science Foundation, in coordination with participating agency officials, should take steps to fully implement leading practices that enhance and sustain collaboration. (Recommendation 5)

AGENCY COMMENTS, THIRD-PARTY VIEWS, AND OUR EVALUATION

We provided a draft of this product to Commerce, DOD, EPA, DOE, DHS, HHS, NASA, NSF, ODNI, OSTP and USDA for comment. Commerce, DOE, NSF, and OSTP generally agreed with the recommendations directed to them.

Commerce, DOE, and NSF provided written comments that are reproduced in appendixes IV, V, and VI, respectively. In expressing concurrence with the recommendations directed to them, these agencies' written comments discussed aspects of the interagency groups' efforts we examined in our report or the agencies' own efforts related to coordination and collaboration.

OSTP's General Counsel provided OSTP's comments by email. In its comments, OSTP stated that it sees value in our recommendation and will implement the recommendation as resources allow. However, OSTP expressed concern about the impact that resource limitations could have on its ability to implement the recommendation. We recognize that OSTP faces certain resource limitations. However, we believe that implementing our recommendation would allow leveraging of limited resources across the agencies participating in a collaborative effort.

In an email from an official with the Office of the Chief Financial Officer in USDA's Agricultural Research Service, USDA provided general comments on our findings and our recommendation pertaining to the

Interagency Working Group on Synthetic Biology. Specifically, USDA concurred that federal support for research and development help drive technological advances and promote U.S. competitiveness. USDA also agreed that the leading practices we discuss in our report can enhance and sustain interagency collaboration, and it expressed support for the implementation of these practices in the Interagency Working Group on Synthetic Biology, consistent with our recommendation.

In addition, Commerce, DHS, DOE, EPA, HHS, NASA, and OSTP provided technical comments, which we incorporated as appropriate. Officials from DOD and ODNI stated via email that they had no comments on the report.

We also provided a draft of this report to a participant who served as moderator in our October 2017 expert meeting on research for transformational technological advances. We requested his views on aspects of the report on which he has expertise and, in particular, the characterization of statements made by experts at our meeting. He provided technical comments, which we incorporated as appropriate.

We are sending copies of this report to the appropriate congressional committees; the Secretaries of Agriculture, Commerce, Defense, Energy, Health and Human Services, and Homeland Security; the Administrators of the Environmental Protection Agency and the National Aeronautics and Space Administration; the Directors of National Intelligence, the National Science Foundation and the Office of Science and Technology Policy; and other interested parties. In addition, this report is available at no charge on the GAO website at http://www.gao.gov.

If you or your staff members have any questions about this report, please contact me at (202) 512-3841 or neumannj@gao.gov. Contact points for our Offices of Congressional Relations and Public Affairs may be found on the last page of this report. Key contributors to this report are listed in appendix VII.

John Neumann

Director, Natural Resources and Environment

Appendix I: Objectives, Scope, and Methodology

The objectives of our review were to (1) describe federal agencies' and nonfederal entities' support for research for transformational technological advances in selected areas, (2) examine federal agencies' coordination on this research, and (3) provide experts' views on considerations for maintaining U.S. competitiveness through transformational technological advances.

For the purposes of this report, we selected quantum computing (a sub-area of quantum information science) and synthetic biology (the intersection of biology and engineering that focuses on the modification or creation of novel biological systems) as examples of research for transformational technological advances. We selected these two areas of research because they: (1) represent enabling or platform technologies, which could lead to other advances, (2) are supported by a mix of federal agencies and nonfederal entities, and (3) represent areas of congressional interest in which we have not recently conducted work.

We conducted this performance audit from November 2016 to September 2018 in accordance with generally accepted government auditing standards. Those standards require that we plan and perform the audit to obtain sufficient, appropriate evidence to provide a reasonable basis for our findings and conclusions based on our audit objectives. We believe that the evidence obtained provides a reasonable basis for our findings and conclusions based on our audit objectives.

Support for Research

To describe federal agencies' and nonfederal entities' support for research for transformational technological advances in quantum computing or synthetic biology we reviewed agency documentation, relevant literature, and our prior work related to federal research efforts. We focused on federal and nonfederal efforts in fiscal years 2016 through the second quarter of fiscal year 2018. For example, we reviewed the

National Science and Technology Council's 2016 report on advancing quantum information science which discusses the state of the research area and federal involvement. We also interviewed officials from 10 agencies and departments that have ongoing work in either quantum computing or synthetic biology, or in some instances, work in both research areas. These agencies were the: Department of Commerce, Department of Defense, Environmental Protection Agency, Department of Energy, Department of Homeland Security, Department of Health and Human Services, National Aeronautics and Space Administration, National Science Foundation (NSF), Office of the Director of National Intelligence, and U.S. Department of Agriculture. We initially selected federal agencies on the basis of those that had total research and development obligations of $500 million or greater in fiscal year 2016 according to NSF's Survey of Federal Funds for Research and Development. Additionally, we included an agency which we learned, through the course of our work, had significant ongoing work in both research areas. We did not seek to develop comprehensive information on federal agencies' efforts to support research in quantum computing and synthetic biology. As a result, federal agencies could have ongoing efforts in these two areas that we do not discuss in our report.

To examine the funding federal agencies provide for quantum computing and synthetic biology research, we requested data on obligations for quantum computing and synthetic biology research for fiscal years 2016 through 2017, information on the type of research funded, and the names of individual studies or projects. We requested funding data from all agencies within our scope but some agencies did not provide such data. We assessed the reliability of the data we obtained by checking for obvious errors in accuracy and completeness and by comparing the data with other sources of funding information, such as agency budget documents, where possible. We determined that the data were sufficiently reliable for reporting an approximate, minimum amount of federal financial assistance obligated for quantum computing and synthetic biology research.

To examine the extent to which nonfederal entities have supported research related to synthetic biology and quantum computing, we interviewed stakeholders from 21 nonfederal entities with experience in the areas of quantum computing, synthetic biology, or federal research more broadly. To collect a range of viewpoints, we selected nonfederal entities from industry, academia, nonprofit organizations, and professional associations. The 21 nonfederal entities we interviewed included:

1) American Chemical Society
2) American Physical Society
3) Arizona State University
4) Georgia Institute of Technology
5) Ginkgo BioWorks
6) Google
7) Harvard University
8) IBM
9) Institute of Electrical and Electronics Engineers
10) Information Technology and Innovation Foundation
11) IonQ
12) Massachusetts Institute of Technology (MIT)[137]
13) Materials Research Society
14) Microsoft
15) National Venture Capital Association
16) Purdue University
17) Science and Technology Policy Institute
18) University of California
19) University of Colorado
20) Virginia Tech
21) Yale University

We also defined the people cited in this report in the following manner:

---

137 We interviewed stakeholders from MIT, including researchers from MIT's Lincoln Laboratory.

1) Experts: individuals who participated in our expert meeting.
2) Stakeholders: academic researchers, industry officials, and representatives of professional organizations who we interviewed. This group does not include agency officials.
3) Agency officials: federal officials we interviewed.

We identified and selected these stakeholders through a literature review and referrals. We conducted a literature review to learn about the current state of each research area as well as to identify relevant stakeholders in the areas of synthetic biology and quantum computing. We then contacted the stakeholders for interviews and asked them for additional references. We interviewed stakeholders both in person and over the phone.

We did not seek to develop comprehensive information on nonfederal efforts to support research in quantum computing and synthetic biology. As a result, we acknowledge that there are nonfederal entities that may have ongoing efforts in these two areas that we do not discuss in our report.

Federal Agencies' Coordination on Research

To examine federal agencies' coordination on quantum computing and synthetic biology research, we identified coordination efforts in fiscal year 2016 through the second quarter of fiscal year 2018 through our review of agency documentation and interviews with federal officials. Additionally, we interviewed officials with the Office of Science and Technology Policy. For ongoing interagency coordination efforts, we compared agencies' efforts with selected leading practices for enhancing and sustaining collaboration.[138] We selected six of the eight practices based on their relevance to the operations of the interagency coordination efforts we

---

[138] GAO, *Results-Oriented Government: Practices That Can Help Enhance and Sustain Collaboration among Federal Agencies*, GAO-06-15 (Washington, D.C.: Oct. 21, 2005).

identified.[139] In this report, and in our past work, we define collaboration broadly as any joint activity that is intended to produce more public value than could be produced when organizations act alone.[140] Through interviews and a data request, we asked agency officials to provide information on their efforts to coordinate quantum computing and synthetic biology research from fiscal year 2016 through the second quarter of fiscal year 2018. For interagency groups related to quantum computing and synthetic biology, we obtained information through June 2018.

Experts' Views

To provide experts' views on considerations for maintaining U.S. competitiveness through transformational technological advances, we convened a meeting of 19 experts on October 12 and 13, 2017, with the assistance of the National Academies of Sciences, Engineering, and Medicine.[141] The experts included current and former federal officials, as well as subject matter experts from industry, academia, nonprofit organizations, and professional associations. About half of the experts were subject matter experts in the areas of quantum computing or synthetic biology, while the other half were experts with broader perspectives on the role of federal and nonfederal entities in supporting research for transformational technological advances. We worked with the National Academies staff to select experts with a range of viewpoints.

Prior to the meeting, we worked with National Academies staff to help ensure balance and to assess potential conflicts of interest among the experts. For example, we asked all participating experts to provide information on (1) whether their immediate family had any investments or assets that could be affected, in a direct and predictable way, by a decision or action based on the information or opinions they would provide to

---

[139] We excluded from our review two leading practices related to reinforcing agency accountability and individual accountability for collaborative efforts.

[140] We also refer to coordination as collaboration in our work.

[141] We planned and convened this meeting of experts with the assistance of the National Academies to better ensure a breadth of expertise; however, we were responsible for all final decisions regarding meeting substance and expert participation. Any conclusions and recommendations in our reports are solely our own.

GAO; (2) whether they or their spouse received any income or hold any organizational positions that could be affected, in a direct and predictable way, by the information or opinions they would provide GAO; and (3) whether there were any other circumstances, not addressed in the two previous questions, that could be reasonably viewed by others as affecting participants' point of view on the topics to be discussed. We received signed responses from all participating experts. Three of the 19 experts reported potential conflicts. We evaluated their statements and determined that they did not have any inappropriate biases when taken in the context of the overall group of experts taking part in the meeting. As a result of these efforts, we determined that the group of 19 experts, overall, was balanced and had no inappropriate biases. However, the views of these experts cannot be generalized to everyone with expertise on research for transformational technological advances; they represent only the views of the experts who participated in our meeting. We list the experts who participated in our meeting in Appendix II.

We divided the 2-day expert meeting into 8 sessions focused on a range of topics, such as the role of federal and nonfederal entities in keeping the United States competitive. Each session featured an opening presentation by two selected experts, followed by open discussion among all meeting participants. At the end of each session, one expert was tasked with highlighting the key themes discussed during that session. We then solicited feedback from the experts to determine whether there were any additional comments they wanted to add to those themes. We recorded and transcribed the meeting to ensure that we accurately captured the experts' statements.

We analyzed the information gathered from the experts by reviewing and conducting a content analysis of the transcript and identifying considerations for maintaining U.S. competitiveness based on categorizing the experts' comments. For purposes of quantifying expert remarks, we refer to a statement from an individual expert as being from one expert, and unless there is significant disagreement in the transcript, we refer to statements from two or more experts as being from experts.

Appendix II: Participants in GAO's Meeting on Research for Transformational Technological Advances

Table 3. Experts Who Participated in GAO's Meeting on Research for Transformational Technological Advances, Held October 12-13, 2017

Expert	Affiliation
Robbie Barbero	Ceres Nanosciences, Inc.
Brad Blakestad	Intelligence Advanced Research Projects Activity
William B. Bonvillian	Massachusetts Institute of Technology
Michael Borrus	XSeed Capital
Tony Dickherber	National Cancer Institute
Drew Endy	Stanford University
Maryann Feldman	University of North Carolina at Chapel Hill
Ralph Gomory	New York University Stern School of Business
Richard Johnson	Global Helix LLC
Christopher Monroe	University of Maryland and IonQ
Pablo Rabinowicz	Department of Energy Office of Science
Marc Salit	National Institute of Standards and Technology
Daniel Sarewitz	Arizona State University
John Sarrao	Los Alamos National Laboratory
Lou Schick	NewWorld Capital
Stephanie Shipp	Biocomplexity Institute of Virginia Tech
Neil Thompson	Massachusetts Institute of Technology
Carl J. Williams	National Institute of Standards and Technology
Robert Wisnieff	IBM TJ Watson Research Center

Source: GAO. | GAO-18-656.

In cases of significant disagreement in the transcript, we refer to statements from two to three experts as being from a few experts, and statements from four to six experts as being from some experts. Before publication and consistent with our quality assurance framework, we provided the experts with a draft of our report and asked them to provide their views on whether our overall characterization of the meeting generally reflected the considerations discussed during the meeting.[142] Of the 18 experts who responded to our request for review, 13 experts agreed that our overall characterization generally reflected the key considerations identified during the meeting, one partially agreed, and one differed with our report's presentation of specific issues regarding synthetic biology.[143] We incorporated feedback experts provided on the draft, as appropriate.

To corroborate statements made by the experts on particular topics, as appropriate, we identified and analyzed studies and reports by agencies, the National Academies, and others that were recommended to us by experts. In addition, we compared the experts' statements to other information provided by agency officials and stakeholders we interviewed.

APPENDIX III: FUNDING/INVESTMENT GAP IN THE MANUFACTURING-INNOVATION PROCESS (CORRESPONDS TO FIGURE 1)

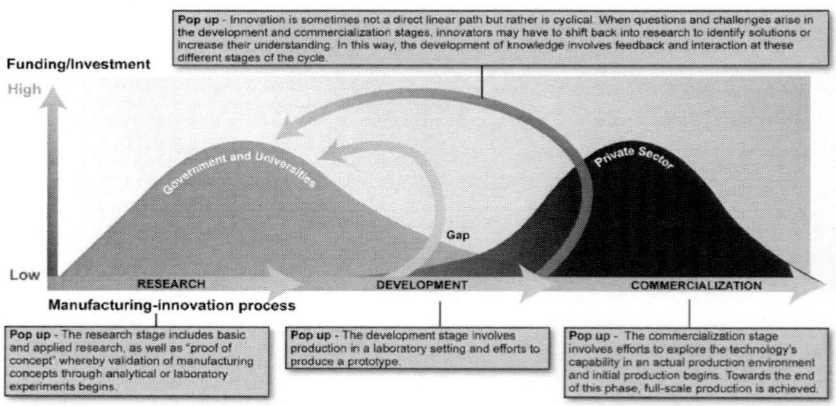

Sourfce: GAO. | GAO-18-656.

Figure 3. Funding/Investment Gap in the Manufacturing-Innovation Process (Corresponds to Figure 1).

---

[142] We also asked experts to provide clarifying information on statements made during the meeting, as necessary.

[143] Three experts did not provide comments on if they agreed or disagreed with our overall characterization of the meeting and one expert did not respond to our requests for feedback.

Figure 3 shows the potential gap during the middle stages of innovation, in which innovators may have difficulty finding financial support. The figure includes a static display of the rollover information included in Figure 1, which is interactive.

APPENDIX IV: COMMENTS FROM THE DEPARTMENT OF COMMERCE

UNITED STATES DEPARTMENT OF COMMERCE
The Secretary of Commerce
Washington, D.C. 20230

September 10, 2018

Mr. John Neumann
Director, Natural Resources and Environment
U.S. Government Accountability Office
441 G Street, NW
Washington, DC 20548

Dear Mr. Neumann:

Thank you for the opportunity to review and comment on the Government Accountability Office's draft report titled *Science and Technology: Considerations for Maintaining U.S. Competitiveness in Quantum Computing, Synthetic Biology, and Other Potentially Transformational Research Areas* (GAO-18-656).

On behalf of the Department of Commerce, I have enclosed our comments on the draft report. We concur with the recommendation to the Department of Commerce and will take steps to implement it.

If you have any questions, please contact MaryAnn Mausser, Department of Commerce Audit Liaison, at (202) 482-8120.

Sincerely,

Wilbur Ross

Wilbur Ross

Enclosure

Department of Commerce's Comments on
GAO Draft Report titled *Science and Technology: Considerations for Maintaining U.S. Competitiveness in Quantum Computing, Synthetic Biology, and Other Potentially Transformational Research Areas*
(GAO-18-656)

The Department of Commerce has reviewed the draft report, and we offer the following comments for the Government Accountability Office's (GAO) consideration.

Comments on Recommendations
The GAO made one recommendation to the Department of Commerce in the report.

- **Recommendation 2:** As the QIS Subcommittee moves forward, the Department of Commerce co-chair, in coordination with other co-chairs and participating agency officials, should take steps to fully implement leading practices that enhance and sustain collaboration.

  Commerce Response: The Department of Commerce concurs with this recommendation.

  The National Institute of Standards and Technology (NIST) of the Department of Commerce is actively engaging with the Department of Energy, the Office of Science and Technology Policy, and the National Science Foundation through the National Science and Technology Council Interagency Subcommittee to ensure that we collaborate closely in guiding and steering our interactions and engagement to maximize the benefit to the U.S. taxpayer. NIST will work to use best practices in this coordination effort. NIST is in the process of establishing an industrial consortium, and invitations to the consortium's inaugural event will include the co-chairs and numerous representatives of the Interagency Subcommittee.

APPENDIX V: COMMENTS FROM THE DEPARTMENT OF ENERGY

Department of Energy
Office of Science
Washington, DC 20585

SEP 10 2018

Mr. John Neumann
Director
Natural Resources and Environment
U.S. Government Accountability Office
441 G Street N.W.
Washington, DC 20548

Dear Mr. Neumann:

We appreciate the opportunity to comment on the draft Government Accountability Office (GAO) report: *Science and Technology: Considerations for Maintaining U.S. Competitiveness in Quantum Computing, Synthetic Biology, and Other Potentially Transformational Research Areas* (GAO-18-656).

Recommendation 3: As the QIS Subcommittee moves forward, the Department of Energy co-chair, in coordination with other co-chairs and participating agency officials, should take steps to fully implement leading practices that enhance and sustain collaboration.

DOE Response: The Department of Energy, including the National Nuclear Security Administration, concurs with the recommendation.

The leadership of the Quantum Information Science (QIS) Subcommittee has developed a strategic overview for QIS that is under review at the participating agencies. The release of strategic overview is planned for September 2018. Upon release, the QIS Subcommittee will proceed with the development of an implementation plan. To inform the implementation plan, the QIS Subcommittee is planning several workshops and summits to discuss leading QIS practices and possible collaborations between industry, academia, and Federal agencies.

Please find an additional technical comment on the report attached. If you have any questions, regarding this response, please contact Barbara Helland, Associate Director for Advanced Scientific Computing Research at (301) 903-7486.

Sincerely,

J. Stephen Binkley
Deputy Director for Science Programs
Office of Science

Printed with soy ink on recycled paper

Appendix VI: Comments from the National Science Foundation

National Science Foundation
Office of the Director

August 31, 2018

John Neumann
Director
Natural Resources and Environment
U.S. Government Accountability Office
441 G Street, NW
Washington, D.C. 20548

Dear Mr. Neumann:

The National Science Foundation (NSF) appreciates the opportunity to review and provide comments on the Government Accountability Office (GAO) draft report, "*Science and Technology: Considerations for Maintaining U.S. Competitiveness in Quantum Computing, Synthetic Biology, and Other Potentially Transformational Research Areas*" (GAO-18-656). Interagency collaboration and coordination is essential to be able to leverage limited federal resources and maximize the impact of strategic investments in research areas such as quantum computing and synthetic biology. Research in these areas have the potential to lead to scientific and technological innovations that will contribute to US economic competitiveness and prosperity.

NSF concurs with the GAO recommendations concerning QIS and synthetic biology. We note that the NSF Director co-chairs the NSTC Committee on Science that oversees the work of both the NSTC QIS subcommittee and the NSTC subcommittee on Biological Sciences, and NSF staff co-chair both of these subcommittees. Thus NSF is well positioned to insure steps are taken to (a) fully implement leading practices that enhance and sustain collaboration in QIS; (b) foster collaboration both across government and with academia and industry to achieve the goals agreed upon by the QIS subcommittee; and (c) enhance and sustain collaboration among the agencies in the area of synthetic biology, a prime goal of the Biological Sciences subcommittee.

On behalf of the NSF staff participating in the GAO review, I would like to acknowledge the GAO team for their diligence and commitment in conducting this review and issuing this report. Please contact Veronica Shelley at (703) 292-4384, if you have any questions or require additional information.

Sincerely,

France A. Córdova
Director

2415 Eisenhower Avenue | Alexandria, VA 22314

GAO Contact and Staff Acknowledgments

GAO Contact

John Neumann, (202) 512-3841 or neumannj@gao.gov

Staff Acknowledgments

In addition to the contact named above, the following individuals made contributions to this report: Christopher Murray (Assistant Director), Angela Miles (Analyst-in-Charge), Justin Fisher, Scott Fletcher, Ashley Grant, Charlotte E. Hinkle, Gwen Kirby, Patricia Moye, Cynthia Norris, Emily Pinto, Tind Shepper Ryen, McKenna Storey, and Walter Vance.

In: Advanced Fossil Energy ...
Editor: Edward Carr
ISBN: 978-1-53615-071-1

Chapter 3

FEDERAL RESEARCH: ADDITIONAL ACTIONS NEEDED TO IMPROVE LICENSING OF PATENTED LABORATORY INVENTIONS*

United States Government Accountability Office

ABBREVIATIONS

AIA	Leahy-Smith America Invents Act
AUTM	Association of University Technology Managers
Commerce	Department of Commerce
CRADA	cooperative research and development agreement
DOD	Department of Defense
DOE	Department of Energy
FLC	Federal Laboratory Consortium
HHS	Department of Health and Human Services
iEdison	Interagency Edison

* This is an edited, reformatted and augmented version of the United States Government Accountability Office Report to the Chairman, Committee on the Judiciary, House of Representatives, Publication No. GAO-18-327, dated June 2018.

LES	Licensing Executives Society
NASA	National Aeronautics and Space Administration
NIH	National Institutes of Health
NIST	National Institute of Standards and Technology
R&D	research and development
Stevenson-Wydler Act	Stevenson-Wydler Technology Innovation Act of 1980
USPTO	United States Patent and Trademark Office
WIPO	World Intellectual Property Organization
WTO	World Trade Organization

WHY GAO DID THIS STUDY

The federal government spends approximately $137 billion annually on research and development—mostly at DOD, DOE, NASA, and NIH—to further agencies' missions, including at federal labs. Multiple laws have directed agencies and labs to encourage commercial use of their inventions, in part by licensing patents, to private sector companies and others that aim to further develop and bring the inventions to market.

GAO was asked to review agency practices for managing inventions developed at federal labs, with a particular focus on patent licensing. This report examines (1) challenges in licensing patents and steps taken to address and report them and (2) information to guide establishing financial terms in patent licenses at DOD, DOE, NASA, and NIH. GAO reviewed relevant literature, laws, and agency documents, including patent licenses from 2014, to match the most recent NIST summary report when the licenses were requested, and GAO interviewed agency officials and knowledgeable stakeholders, including organizations that assist federal labs in licensing patents.

WHAT GAO RECOMMENDS

GAO is making seven recommendations, including that Commerce instruct NIST to fully report the range of challenges in federal patent licensing in its annual reports to Congress and facilitate information sharing among agencies. Commerce, DOD, DOE, NASA, and NIH generally agreed with GAO's recommendations and are taking steps to implement them.

WHAT GAO FOUND

Federal agency and laboratory (lab) officials identified challenges in licensing patents across the federal government, and agencies have taken some steps to address and report them. Patent licensing is a technology transfer activity that allows, for example, federal inventions to be legally transferred to the private sector for commercial use.

Specifically, officials at the Departments of Defense (DOD) and Energy (DOE), National Aeronautics and Space Administration (NASA), and National Institutes of Health (NIH), as well as external stakeholders, noted challenges in having researchers identify potentially patentable inventions. DOD, DOE, and NIH officials also cited having inadequate internal systems to keep track of inventions developed in the labs. In addition, several stakeholders stated that licensing patented inventions can be lengthy and bureaucratic, which may deter companies from licensing. The agencies reported taking steps to address these challenges, such as implementing model license agreements across labs to expedite the process.

The Department of Commerce has delegated to its National Institute of Standards and Technology (NIST) to annually report agencies' technology transfer activities, including patent licensing. Although NIST has reported some challenges, it has not fully reported the range of challenges identified by agency and lab officials and stakeholders. NIST officials stated that they were generally aware of the challenges but had not considered including them to a greater degree in their annual reports to Congress. By fully reporting the range of challenges in federal patent licensing, NIST has the opportunity to further ensure that Congress is more aware of challenges that limit agencies' efforts and ways for potentially addressing those challenges.

Selected Challenges in Licensing Federal Inventions and Steps Taken to Address Them

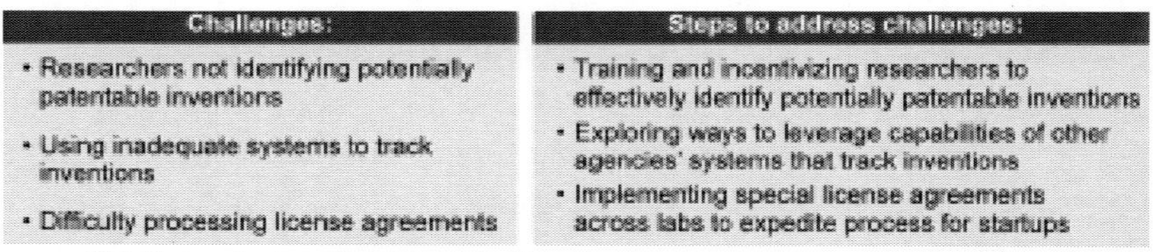

Source: GAO analysis of Relevant literature and agency an external stakeholder interviews. | GAO-18-327.

Federal agencies and labs have limited information to guide officials when establishing the financial terms of patent licenses. For example, while federal labs can use comparable licenses to help establish financial terms, their access to information on comparable licenses from other labs varies, and such information is not formally shared among the agencies. Based on its established interagency role, NIST is best positioned to assist agencies in sharing information on comparable licenses, in accordance with leading practices for interagency collaboration. By doing so, NIST would provide federal agencies and labs with useful information that can help them better establish financial terms and successfully license inventions.

* * *

June 19, 2018

The Honorable Bob Goodlatte

Chairman

Committee on the Judiciary House of Representatives

Dear Mr. Chairman:

The federal government spends approximately $137 billion annually on research and development (R&D) to help further agencies' missions, including at federal laboratories (lab).[1] The intellectual property generated by this research—including inventions—has application beyond federal agencies' immediate, mission-related goals if the private sector can bring the inventions to market.[2] For example, government research has led to new products and processes for the commercial marketplace, including antibiotics, plastics, airplanes, computers, microwaves, and bioengineered drugs. However, a 2013 Office of Science and Technology Policy report raised concerns that only a small portion of the inventions arising from government research have been commercialized by the private sector, and that the United States is potentially missing critical opportunities to improve the nation's standard of living, create new jobs, maintain international competitiveness, and enhance the overall economy, among other things.[3]

Technology transfer is the process of transferring scientific findings from one organization to another for the purpose of further development

---

[1] We use the term lab in a broad sense to include all federally funded labs and R&D centers. Annual federal R&D spending (obligations) averaged $137 billion for fiscal years 2015 to 2017.

[2] Section 4 of the Stevenson-Wydler Technology Innovation Act of 1980, as amended, defines "invention" as any invention or discovery that is or may be patentable or otherwise protected under title 35, United States Code, or any novel variety of plant that is or may be protectable under the Plant Variety Protection Act (7 U.S.C. § 2321 et seq.). Pub. L. No. 96-480 § 4 (codified as amended at 15 U.S.C. § 3703(7)). Intellectual property is defined as creative works or ideas embodied in a form that can be shared or can enable others to re-create, emulate, or manufacture them. Patents, trademarks, copyrights, and trade secrets are forms of intellectual property protection.

[3] White House Office of Science and Technology Policy and the National Institutes of Health, *White House Lab-to-Market Inter-Agency Summit: Recommendations from the National Expert Panel* (Washington, D.C.: May 20, 2013).

and commercialization.[4] A series of federal laws and executive orders spanning nearly 40 years have directed federal agencies to enhance their labs' beneficial impact on society through technology transfer. For example, the Stevenson-Wydler Technology Innovation Act of 1980 (Stevenson-Wydler Act) states that its purpose is to improve the economic, environmental, and social well-being of the United States by, among other things, stimulating improved utilization of federally funded technology developments by nonfederal entities.[5] Accordingly, the act provides that the federal government, where appropriate, shall transfer federally owned or originated technology to state and local governments and to the private sector.

From fiscal years 2015 to 2017, four federal agencies consistently had the highest annual federal R&D spending: the Department of Defense (DOD), the Department of Energy (DOE), the National Aeronautics and Space Administration (NASA), and the National Institutes of Health (NIH) within the Department of Health and Human Services (HHS).[6] Together, these four agencies accounted for nearly 90 percent of total federal R&D spending on average, as shown in figure 1.

One way federal agencies transfer technology is by licensing patents on inventions generated by R&D at federal labs.[7] This licensing aims to put

---

[4] In some cases, technology transfer involves the transfer of legal rights, such as licensing a government-owned patent to a private sector entity. Technology transfer also includes collaboration between private companies and federal labs, for example, in the testing of advanced batteries. In other instances, technology transfer involves the informal transmission of information, knowledge, and skills through person-to-person or organization-to-organization interaction. Commercialization is the process of developing marketable products or services and producing and delivering products or services for sale.

[5] Pub. L. No. 96-480, § 3, 94 Stat. 2311, 2312 (codified as amended at 15 U.S.C. § 3702).

[6] We use the term spending to refer to agency obligations on R&D, as reported by the National Science Foundation, National Center for Science and Engineering Statistics, *Survey of Federal Funds for Research and Development, FYs 2015–17* (Arlington, Va.: Apr. 5, 2017). Total R&D spending includes all direct, incidental, or related costs, for both intramural and extramural R&D, and does not directly correspond to the R&D spending used to develop, patent, and license inventions at federal labs.

[7] A patent is an exclusive right granted for a fixed period of time to someone who invents or discovers (1) a new and useful process, machine, manufacture, or composition of matter or (2) any new and useful improvement of such items. A patent owner can prevent others from making, using, selling, or offering for sale the patented invention in the United States, or

federal inventions in the hands of those equipped to further develop the inventions into marketable products accessible to the public or to find other practical applications for them.[8] Examples of inventions developed at DOD, DOE, NASA, and NIH labs can be found in appendix I. Numerous statutory and regulatory requirements have been established to help ensure that agencies commercialize inventions arising from R&D at federal labs. The Department of Commerce (Commerce) is responsible for government-wide functions related to patenting, including granting patents through the United States Patent and Trademark Office (USPTO). Commerce is also responsible for functions related to licensing of federally owned inventions, including reporting federal agencies' technology transfer activities to Congress and issuing regulations, both of which it delegated to the National Institute of Standards and Technology (NIST).

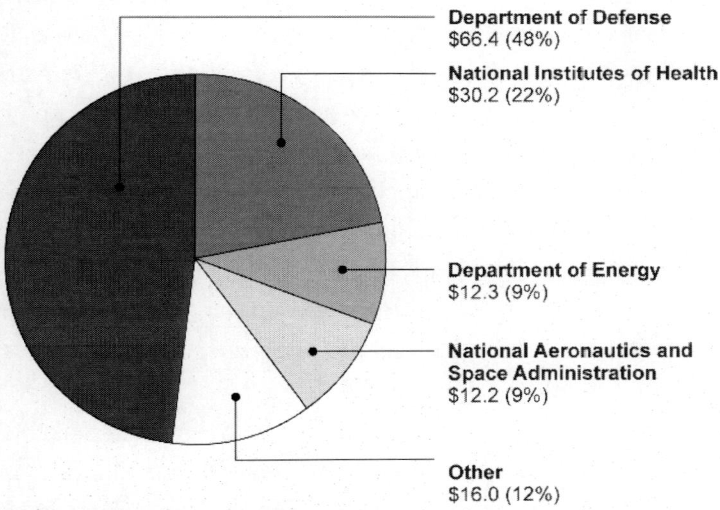

Source: GAO analysis of National Science Foundtion data. | GAO-18-327.

Figure 1. Federal Research and Development Spending by Agency, Average for Fiscal Years 2015 to 2017 (Dollars in Billions).

---

importing it into the United States without authorization. A patent owner can license or assign the patent rights.

[8] According to Department of Commerce officials, "practical application" is important in the law because some inventions that involve, for example, defense technologies are not for commercial use per se, other than for being sold to DOD.

Over the years, we and others have reported on a range of challenges that agencies face in commercializing inventions arising from R&D at federal labs.[9] You asked us to review agency practices for managing intellectual property developed at federal labs, with a particular focus on the licensing of patented inventions (patent licensing) to nonfederal parties that might use the patented inventions to manufacture products. This report examines, for DOD, DOE, NASA, and NIH and their labs, (1) challenges that federal labs face in patent licensing, steps taken to address those challenges, and the extent to which NIST has reported them and (2) the extent to which federal agencies and labs have information on processes, goals, and comparable licenses to guide establishing patent license financial terms.

For both objectives, we reviewed statutes and regulations applicable to patent licensing to describe the legal framework governing federal patent licensing and license financial terms. To obtain views on patent licensing practices across the federal government, we conducted 60 semistructured interviews, including 38 with agency and lab officials. We interviewed officials from DOD, DOE, NASA, and NIH, as well as from nine federal labs across these agencies. We interviewed officials at multiple labs at each agency and selected them to obtain a variety of perspectives on patent licensing, including variation in whether they were contractor-operated (two labs) or government-operated labs (seven labs); the volume and type of licensing activity; and other agency-specific considerations, such as whether they use a service center model.[10]

In addition, we conducted 22 interviews with external stakeholders—including academic researchers (4), partnership intermediaries (11) (i.e., organizations that assist federal labs and businesses in licensing federal patents), industry representatives and companies (3), and professional trade

---

[9] GAO, Technology Transfer: Federal Laboratory Consortium Should Increase Communication with Potential Customers to Improve Initiatives, GAO-15-127 (Washington, D.C.: Oct. 3, 2014); Defense Technology Development: Technology Transition Programs Support Military Users, but Opportunities Exist to Improve Measurement of Outcomes, GAO-13-286 (Washington, D.C.: Mar. 7, 2013); and Technology Transfer: Clearer Priorities and Greater Use of Innovative Approaches Could Increase the Effectiveness of Technology Transfer at Department of Energy Laboratories, GAO-09-548 (Washington, D.C.: June 16, 2009).

[10] A service center model uses a central technology transfer office to serve multiple labs.

organizations and universities (4)—that were knowledgeable about federal patent licensing practices. Partnership intermediaries were selected based on lists of partnership intermediaries associated with the selected labs and agencies that the agencies submitted and referrals from other partnership intermediaries. To select other stakeholders, we used snowball sampling based on referrals obtained from prior engagements, our review of academic literature, and referrals from stakeholder interviews during the engagement. We also interviewed officials from NIST and USPTO. The information we obtained from these interviews is not generalizable to all agency and lab officials and external stakeholders, but we determined that our selection of interviewees was appropriate to obtain varied perspectives on the patent licensing process and related challenges and that the selection would generate valid and reliable evidence to support our work.

To understand the level of patent licensing activity and the terms in patent licenses at federal labs, we requested data from each agency on licenses that were active as of the end of fiscal year 2014.[11] To assess the reliability of the data on active patent license agreements, we asked agency and lab officials questions about the accuracy and completeness of the data and asked them to confirm specific information. Based on these steps, we found the data to be sufficiently reliable for understanding the overall level of licensing activity in the labs, for identifying licenses that became effective in fiscal year 2014, and as a source of information for those licenses. We also requested from each agency and reviewed the subset of licenses that became effective in fiscal year 2014.

To address the first objective, we first conducted a literature search.[12] We broadened our search beyond articles published in peer-reviewed journals to identify studies, such as dissertations, conference proceedings, and organizational studies issued by research institutes or studies issued by government agencies.[13] For example, we conducted both subject and

---

[11] We requested data for licenses active at the end of fiscal year 2014 to match the data in the NIST fiscal year 2014 summary report, which was the most recent when the data were requested.

[12] See the selected bibliography following the appendixes.

[13] "Organizational studies" refers to those studies published by nongovernmental organizations, such as the Heritage Foundation and the Science and Technology Policy Institute.

keyword searches related to patent licensing in various databases.[14] We performed these searches and identified 23 studies from 2000 to 2016—2016 was the year for which the most recent information was available at the time of our review.

We reviewed relevant agency documentation and interviewed the agency and lab officials and external stakeholders identified above to obtain their perspectives on areas of the patent licensing process identified from our review of relevant federal statutes and regulations.[15] Using information obtained from the interviews, we conducted a content analysis of responses. Challenges in each area were identified from the content analysis based on the frequency and consistency of responses from agency and lab officials and stakeholders. We also examined other relevant agency documentation, including NIST's fiscal year summary reports to Congress from fiscal years 2013 through 2015—the most recent reports available—to determine the extent to which challenges in federal patent licensing had been reported.

To address the second objective, we reviewed relevant economic literature on establishing the financial terms of patent licenses and other related licensing terms. From the literature, we identified economic principles that apply to structuring financial terms to promote the commercial use of inventions. We reviewed all relevant documentation describing licensing practices at the agency and lab levels that DOD, DOE, NASA, and NIH provided. This included documentation describing, among other things, processes and goals for establishing license terms, including financial terms. We reviewed each document to identify (1) factors considered in establishing financial terms; (2) data sources used in the process; (3) methods, such as calculations or guidelines, for setting financial terms; (4) guidance on when or how to use various types of financial terms; and (5) goals applicable to financial terms. We also reviewed narratives

---

[14] These include OCLC WorldCat, Web of Science, DIALOG (SciSearch, Social SciSearch, MEDLINE, Chembase), Scopus, ProQuest, SSRN, NBER, and LEXIS.

[15] The basis for the areas of the federal patent licensing process we chose was our review of part 404 of title 37 of the *U.S. Code of Federal Regulations* and the Federal Laboratory Consortium's *Technology Transfer Desk Reference: A Comprehensive Guide to Technology Transfer*, 6th ed. (Cherry Hill, N.J.: 2013).

and written responses describing lab practices for establishing financial terms in patent licenses that the agencies and labs provided in response to our requests. To identify patents at the labs under the four agencies we reviewed and the extent to which they patent in similar technology fields, we searched the assignee field in USPTO's PatentsView database for patents issued since 2000 using search terms associated with the four agencies and their contractor labs, if applicable.[16] The list of assignees for the patents from the initial search was reviewed to eliminate patents that may not have been assigned to the four agencies. In total, the search yielded 20,612 patents assigned to the four agencies.[17] To examine similarities across technology areas, we used PatentsView data on technology sectors and fields classified according to the World Intellectual Property Organization (WIPO) international classification system for patents.[18]

We conducted this performance audit from June 2016 to June 2018 in accordance with generally accepted government auditing standards. Those standards require that we plan and perform the audit to obtain sufficient, appropriate evidence to provide a reasonable basis for our findings and conclusions based on our audit objectives. We believe that the evidence obtained provides a reasonable basis for our findings and conclusions based on our audit objectives.

---

16 These search terms included the name of the agency, the lab, and the lab contractor (if applicable).

17 The 20,612 patents should not be considered an exhaustive list of patents assigned to these agencies or their contractor labs. While many patents associated with each agency were found, some contractor labs may be underrepresented because of challenges in separating out patents associated with the lab from those associated with a larger organization. However, additional patents would only add to the potential for overlap in technology fields.

18 WIPO is a specialized agency of the United Nations, with 191 member states. The organization's mission is to lead the development of a balanced and effective international intellectual property system that enables innovation and creativity for the benefit of all. According to the organization, the International Patent Classification, established by the Strasbourg Agreement of 1971, provides for a hierarchical system of language-independent symbols for classifying patents and utility models according to the different areas of technology to which they pertain.

BACKGROUND

This section outlines the legal framework under which agencies and federal labs license patents and general stages of the patent licensing process.

Legal Framework for Patent Licensing

Prior to 1980, federal agencies generally retained title to any inventions developed through federally funded research—whether extramural, that is, conducted by universities and contractors, or intramural, conducted by federal agencies in their own facilities. By the late 1970s, there was increasing debate in Congress over ways to allow the private and public sectors better access to federally owned inventions by, among other things, creating a uniform policy for those seeking to license inventions developed in federal labs. In the 1980s, Congress began passing a series of key laws that have provided the foundation for federal technology transfer activities, including patenting and licensing inventions that are developed in federal labs and funded by federal dollars. One of the first technology transfer laws, the Stevenson-Wydler Act, established technology transfer as a federal policy and required federal labs to set up Offices of Research and Technology Applications (which, for our purposes, we refer to as technology transfer offices) and devote budget and personnel resources to promoting the transfer of federal technologies to the private sector.[19] In 1980, another key law, the Bayh-Dole Act allowed not-for-profit corporations, including universities, and small businesses to retain title to their federally funded inventions.[20] In 1984, through

[19] Pub. L. No. 96-480, 94 Stat. 2311 (codified as amended at 15 U.S.C. §§ 3701-3715, 3719-23).

[20] Patent and Trademark Law Amendments Act, Pub. L. No. 96-517, 94 Stat. 3015 § (codified in 35 U.S.C. § 200-211), commonly referred to as the Bayh-Dole Act. In 1987, by Executive Order 12,591, 52 Fed. Reg. 13,414 (Apr. 22, 1987), and Executive Order 12,618, 52 Fed. Reg. 48,661 (Dec. 24, 1987). The policy expressed in the Bayh-Dole Act directed at not-for-profit corporations, universities, and small businesses was expanded to apply to all recipients of federal funds, regardless of size, including large businesses.

amendments made to the Bayh-Dole Act, Commerce became responsible for issuing regulations to implement the act.[21]

The Stevenson-Wydler Act was amended by the Federal Technology Transfer Act of 1986, which (1) established the Federal Laboratory Consortium (FLC); (2) required that technology transfer efforts be considered positively in employee performance evaluations; and (3) empowered federal agencies to permit the directors of government-owned, government-operated labs to enter into cooperative research and development agreements (CRADA) and to negotiate license agreements for inventions created in the labs.[22]

The FLC began largely as a forum for the education, training, and networking of federal technology transfer officials to promote the integration of technical knowledge that federal departments and agencies developed into the U.S. economy. Over time, the FLC's role would include serving as a clearinghouse—a central point for collecting and disseminating information—for federal technologies and assisting outside entities in identifying available federal technology. Within Commerce, NIST is the designated host and financial administrator of the FLC.

Additional laws were adopted to help further the development of federally owned inventions for commercial use. Among them was the National Competitiveness Technology Transfer Act of 1989, which directed federal agencies to propose, for inclusion in contracts, provisions to establish technology transfer as a mission of government-owned, contractor-operated labs and permitted those labs, under certain circumstances, to enter into CRADAs.[23] In addition, the Technology

---

21 Trademark Clarification Act of 1984, Pub. L. No. 98-620, § 501(10) (amending 35 U.S.C. § 206). Prior to the 1984 amendments, the Office of Federal Procurement Policy was authorized to develop these regulations.

22 Pub. L. No. 99-502, § 2, 100 Stat. 1785, 1785, (amending Pub. L. No. 96-480, adding § 12, codified as amended at 15 U.S.C.

23 Pub. L. No. 101-189, div. C, tit. XXXI, pt. C, § 3133(d)(1)(A), 103 Stat. 1352, 1678.

Transfer Commercialization Act of 2000 required Commerce to provide Congress with summary reports on agencies' patent licensing and other technology transfer activities.[24] Since 2007, Commerce has delegated to NIST the role of providing to Congress an annual report summarizing technology transfer at federal agencies.[25] NIST's role as the lead in an interagency collaborative effort in federal technology transfer grew further when Commerce delegated to the agency the additional responsibility of coordinating the Interagency Working Group for Technology Transfer.[26] Commerce also has delegated to NIST its authority to promulgate implementing regulations pertaining to patenting and licensing at federal labs. In 2011, Congress passed the Leahy-Smith America Invents Act (AIA) that further affected technology transfer activities by federal labs through comprehensive changes made to the U.S. patent system.[27]

Federal Labs

Federal labs are typically managed under either a government-operated or a contractor-operated model. Commerce regulations prescribe the terms,

---

[24] Pub. L. No. 106-404 § 10(a)(3), 114 Stat. 1742, 1747 (codified at 15 U.S.C. 3710(g)(2)). § 3710a(a)).

[25] In 2007, the America Creating Opportunities to Meaningfully Promote Excellence in Technology Education and Science (COMPETES) Act eliminated Commerce's Technology Administration, which had been responsible for collecting information on all federal agencies' technology transfer activities and submitting the information to Congress and the Office of Management and Budget. Pub. L. No. 110-69 § 3002(a)(1), 121 Stat. 572, 586 (2007).

[26] Through Department Organization Order 30-2A, Commerce delegated to NIST the responsibility of coordinating the Interagency Working Group for Technology Transfer, which was established in 1987 by Executive Order 12591. The working group includes heads of representative agencies and their directors, or designees, of federal labs who identify and disseminate approaches in technology transfer to the agencies and labs.

[27] Pub. L. No.112-29, 125 Stat. 284 (2011). Signed into law on September 16, 2011, the act implemented changes to the patent system that included giving the inventor with the earliest filed patent application—not the earliest inventor—the right to the patent, creating within USPTO the Patent Trial and Appeal Board that replaced the Board of Patent Appeals and Interferences, and establishing three new proceedings for challenging issued patents. Pub. L. No.112-29 §§ 3, 6, 7, 18.

conditions, and procedures that government-operated labs are to use to license their inventions for commercial use or other practical applications.[28] Government-operated labs are usually owned or leased by the federal government and are predominantly staffed by federal employees. Contractor-operated labs, on the other hand, operate facilities and equipment that are owned by the federal government, but the staff is employed by a private or nonprofit contractor that operates the lab under a contract with the federal government.[29]

Contractor-operated labs typically license their technologies under the authority of the Bayh-Dole Act, applicable regulations, and their contracts, which generally give contractor-operated labs more flexibility in licensing their technologies.[30] Contractors that manage and operate labs include universities, private companies, nonprofit organizations, or consortia thereof. As discussed below, whether a lab is government-operated or contractor-operated will affect how that lab licenses inventions because each type operates under a different set of licensing regulations and requirements.

The Federal Licensing Process

The pathway of an invention from lab development to commercial product can end at any point, and products may not always reach, or find success in, the marketplace. Figure 2 shows the seven general areas of the patent licensing process at federal labs.

---

[28] The regulations define "practical application" to mean "to manufacture in the case of a composition or product, to practice in the case of a process or method, or to operate in the case of a machine or system; and, in each case, under such conditions as to establish that the invention is being utilized and that its benefits are to the extent permitted by law or Government regulations available to the public on reasonable terms." 7 C.F.R. § 404.3(d).

[29] Federal Laboratory Consortium, *FLC Technology Transfer Desk Reference*.

[30] For example, contractor-operated labs are not required to obtain a commercialization plan prior to licensing. A commercialization plan is a detailed description of an applicant's plan for development of the invention, marketing of the invention, or both.

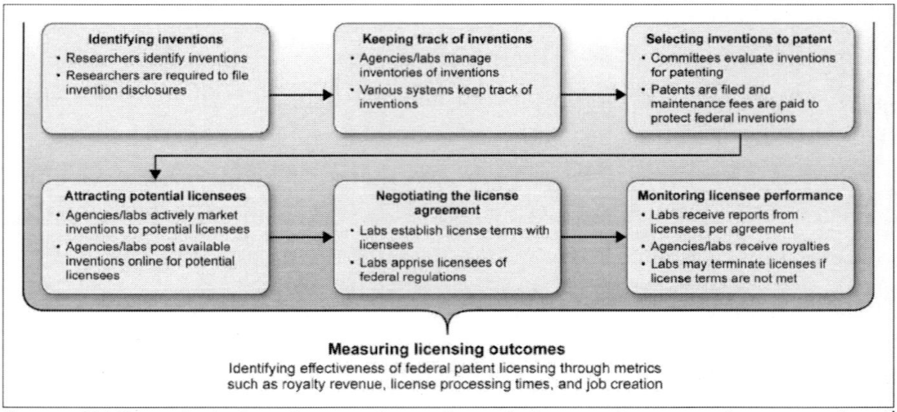

Source:GAO analysis based on review of regulations and agency documentation. | GAO-8-327.

Figure 2. Federal Patent Licensing Process.

The pathway of an invention from lab development to commercial product can end at any point, and products may not always reach, or find success in, the marketplace. Figure 2 shows the seven general areas of the patent licensing process at federal labs.

Identifying Inventions

The patent licensing process begins with researchers identifying inventions—a process that primarily relies on researchers disclosing their inventions to lab officials, mostly through the lab director or directly to an agency's technology transfer office. Various laws and regulations establish a uniform policy for determining who holds the rights to government employees' inventions.[31] Some government-operated labs allow or encourage researchers to publish their research, including research describing inventions, for public dissemination, such as in research journals. Contractor-operated labs are required to disclose inventions to the agency within 2 months after notifying contractor personnel responsible for patent licensing activities. Labs must then decide within 2 years after

[31] See, for example, 35 U.S.C. § 202; 37 C.F.R. pt. 501.

the disclosure whether to retain title to the invention.[32] The contract then must file its initial patent application on the invention to which it elects to retain title within one year after election of title. If the contractor-operated lab does not disclose the invention or elect to retain title within the times specified in the law and regulations, it will convey title to the invention to the funding agency upon written request.[33].

Keeping Track of Inventions

Once an invention has been identified and disclosed, federal agencies and labs keep track of the invention. How they do so varies in degree of automation and centralization. For example, systems that keep track of lab inventions can range from spreadsheets to automated software that tracks all patent licensing and other technology transfer activities. Also, such systems can be centralized, with oversight at the agency level, or decentralized, with independent oversight at the lab level—which is generally the case at contractor-operated labs. Some contractor-operated labs manage their federally funded inventions through the Interagency Edison (iEdison) reporting system, which is owned and managed by NIH.[34]

Selecting Inventions to Patent

Before applying for patent protection through USPTO, agency and lab officials review the invention—often using evaluation committees and patent attorneys—to consider a number of factors, including whether it is patentable, it furthers the lab's mission, and patenting the invention is likely to bring it to commercial use or practical application. The agency must file a patent application within 1 year of the first publication, public use, sale, or offer for sale of the invention or lose U.S. patent rights to that

---

[32] 37 C.F.R. § 401.14.

[33] According to USPTO officials, the same is applicable to the failure to file or prosecute patent applications, or maintain or defend issued patents in administrative proceedings, in any given country.

[34] iEdison allows government grantees and contractors to report federally funded subject inventions, patents, and utilization data via the web to the government agency that issued the funding award. More than 30 U.S. federal funding agency offices use iEdison.

invention.[35] Not all patents will be licensed out to companies for a variety of reasons, including national security considerations. The average time from filing to issuing a patent, or when an application is abandoned, is about 2 years, according to USPTO. Patent applications are often rejected, modified, and refiled, and various fees are associated with filing and prosecuting a patent application. However, according to USPTO, patent maintenance fees that allow federal labs to maintain their patents in force are among the most significant fees.[36]

Attracting Potential Licensees

Agencies and labs use a variety of methods to attract potential licensees, including those from industry, universities, and nonprofits.[37] For example, agencies may post their inventory of patented inventions online, publish them in academic journals, or highlight them at public events. Agencies and labs actively engage with the private sector by, for example, attending conferences where companies can network with federal researchers and federal technology transfer officials. In addition, technology transfer offices often work with partnership intermediaries—such as local or state entities and nonprofit organizations—to support their efforts, including reaching out to potential licensees.[38] Labs have other mechanisms to help attract potential licensees to further develop their inventions. For example, CRADAs can help facilitate licensing or the transfer of knowledge from a lab to a licensee, and new inventions that arise under a CRADA are typically made available to the partner via an option to license.

---

[35] Contractor-operated labs typically apply for patents on inventions within 1 year of the election of title, while providing additional reports to the agency on their progress in commercializing the inventions.

[36] Agencies are to pay maintenance fees to USPTO at 3.5, 7.5, and 11.5 years after the date of issue, with the associated fees increasing over the life of the patent. USPTO officials stated that a patent for an invention, known as a utility patent, is granted for a term that usually ends 20 years from the date the federal lab or inventor first filed for the patent, subject to the payment of appropriate maintenance fees.

[37] A licensee, in this context, is an entity that gains a legal license to exercise a patent for an invention from its owner (i.e., a federal lab or contractor).

[38] Partnership intermediaries are authorized by 15 U.S.C. § 3715 to help federal agencies or laboratories with technology transfer to increase the likelihood of success.

Negotiating the License Agreement

The technology transfer offices and legal counsel are generally responsible for crafting and negotiating the terms of the patent license, sometimes with input from other lab officials. Negotiations are often an iterative process in which both the lab and the licensee request adjustments to the terms of the license. Laws and regulations specify some terms that government-operated labs must include in their licenses. Among others, a typical license includes terms related to (1) financial compensation (if applicable), (2) the degree of exclusivity of the license, (3) the U.S. manufacturing requirement, (4) retained rights for the government, (5) termination of the license, and (6) enforcement of licenses.

Financial terms may include up-front fees; minimum payments; royalties, usually based on sales; and milestone payments, among others. Federal labs typically establish financial terms on a case-by-case basis that are tailored to the specifics of the technology, licensee, and market conditions.[39] License agreements may be nonexclusive, partially exclusive, or fully exclusive, and may be limited to some fields of the invention's use or to specific geographic areas.[40]

Government-operated labs must publicly announce their intent to grant an exclusive license for at least 15 days. After this period, comments and objections are considered. Negotiations then begin with the proposed licensee or, if the licensee has changed, another public announcement of the new licensee may be required. Government-operated labs are required to obtain a commercialization plan from a potential licensee regardless of the degree of exclusivity. Contractor-operated labs, which typically retain title to their inventions under the authority of the Bayh-Dole Act, are not subject to the requirement to obtain a commercialization plan from a prospective licensee before granting a license; however, they are subject to

---

39 There are certain limited exceptions for special licensing programs—often targeted at start-ups—in which the terms of the license may be preset. For the purposes of our report, start-ups are newly created small businesses.

40 For the purposes of this report, an exclusive license grants the licensee the sole right to use, manufacture, and sell a patented invention, and a partially exclusive license is an exclusive license that is limited to some fields of the invention's use or to some specified geographic areas.

requirements specified in their contracts regarding patent licensing. In addition, they are not subject to the same notification requirements as government-operated labs.[41]

The law also contains some other provisions pertaining to patent licenses originating from federal labs. For example, the law generally gives preference to small businesses that are capable of bringing the invention to practical application. There is a general preference for products that incorporate federal inventions to be manufactured substantially in the United States; however, on a case-by-case basis, agencies may waive this requirement. Applicable law also reserves certain rights for the government to protect the public's interests in federally funded inventions. For example, the government retains a royalty-free license to use inventions that are contractor owned or that are licensed exclusively. In addition, the Bayh-Dole Act provides the government march-in authority when certain statutory conditions have been met. Under this authority, an agency may grant a license to an invention developed with federal funding even if the invention is exclusively licensed to another party if, for example, it determines that such action is needed to alleviate public health or safety needs which are not reasonably satisfied by the contractor, assignee, or their licensee.[42] A federal lab can also terminate a license when the licensee is not meeting its commitment to achieve practical application of the invention. The lab can also, through the license, grant permission to a licensee to pursue patent infringement cases.

Monitoring Licensee Performance

Federal license agreements generally require licensees to report periodically on their commercialization.[43] For instance, labs generally put specific monitoring requirements in the license agreements, including milestones and reporting requirements. Through the agreements, government-operated labs have the right to terminate or modify licenses if

---

[41] The applicable acquisition regulations for patents and licensing at DOE contractor-operated labs are found at 48 C.F.R. 970.5227-3, 48 C.F.R. 970.5227-10, and 48 C.F.R. 970.5227-12.

[42] 35 U.S.C. § 203(a)(1), (2); see also 37 C.F.R. §401.14(j).

[43] 35 U.S.C. § 209(d)(2); see also 37 C.F.R. § 404.5(b)(6).

certain requirements are not met.[44] Government-operated labs must submit written notices to the licensees and any sublicensees of their intentions to modify or terminate licenses, and allow 30 days for the licensees or sublicensees to remedy any breach of the licenses or show cause why the licenses should not be modified or terminated.[45] Contractor-operated labs also monitor licensee performance in much the same way; however, they are subject to a different set of regulations.

Measuring Licensing Outcomes

Federal labs are responsible for measuring the outcomes of their activities in all areas of the patent licensing process by developing metrics and evaluation methods. Measuring licensing outcomes help labs assess the effectiveness of their patent licensing efforts. Soon after the passage of AIA, President Obama issued a memorandum in October 2011 to the heads of executive departments and agencies calling for, among other things, (1) developing strategies to increase the usefulness and accessibility of information about federal technology transfer opportunities; (2) listing all publicly available, federally owned inventions on a public government database; and (3) improving and expanding its collecting of metrics for Commerce's annual technology transfer summary report.[46]

Technology Transfer and Agency Mission

Federal law states that it is Congress's policy and objective to use the patent system to promote the commercialization and public availability of inventions, and that technology transfer, including federal patent licensing, is the responsibility of each laboratory science and engineering professional. No single federal agency is responsible for managing technology transfer activities government-wide. Rather, each federal

---

[44] 35 U.S.C. § 209(d)(3); see also 37 C.F.R. § 404.5(b)(8).

[45] 37 C.F.R. § 404.10.

[46] Memorandum on Accelerating Technology Transfer and Commercialization of Federal Research in Support of High-Growth Businesses, DCPD 201100803 (Oct. 28, 2011).

agency involved in technology transfer designs its own program to meet technology transfer objectives, consistent with its other mission responsibilities.

Federal Labs and Stakeholders Identified Challenges in Patent Licensing, and Agencies Have Taken Some Steps to Address Them, but NIST Has Not Fully Reported Such Challenges

Federal agency and lab officials and external stakeholders have identified challenges across the federal patent licensing process, but NIST has not fully reported such challenges. Specifically, DOD, DOE, NASA, and NIH officials at the agency and lab levels, as well as external stakeholders, cited challenges related to all seven areas of the patent licensing process. In addition, officials and stakeholders cited challenges in one area that cuts across the entire process: prioritizing patent licensing as part of agencies' missions. In its annual reports to Congress on federal labs' performance in patent licensing activities, NIST has discussed some challenges identified by agency and lab officials and external stakeholders but has not fully reported on the range of challenges they have experienced.

Federal Labs and Stakeholders Identified Challenges across the Patent Licensing Process, and Agencies Have Taken Some Steps to Address These Challenges

DOD, DOE, NASA, and NIH officials at the agency and lab levels, as well as external stakeholders, identified challenges in all seven areas of the patent licensing process, including identifying inventions, keeping track of inventions, and negotiating license agreements.[47] They also cited challenges in prioritizing patent licensing as part of agencies' missions. Based on our analysis of relevant literature and on interviews with external stakeholders, many of these challenges are occurring government-wide. DOD, DOE, NASA, and NIH have taken some steps to address the challenges in each area of the patent licensing process.

Challenges in Implementing the Patent Licensing Process

DOD, DOE, NASA, and NIH officials at the agency and lab levels, as well as external stakeholders, identified challenges in all seven areas of the patent licensing process, including not identifying inventions, keeping track of inventions in inadequate systems, and difficulty negotiating license agreements. For example, several DOD, DOE, NASA, and NIH officials stated that some researchers do not have adequate training in identifying potentially patentable inventions. When a federal researcher does not disclose to lab officials an invention developed in a federal lab, the opportunity to assess the invention's potential for commercial use may be lost. Federal officials cited various reasons why researchers do not disclose inventions. Navy officials, for example, stated that researchers are often intimidated by the overall invention disclosure process and tend to focus on their research rather than consider what could be patentable. Officials at one NASA lab noted that they have come across a few contractor employees who do not see the benefit of filing invention disclosures, and sometimes researchers are too busy to engage in the patenting process.

Our analysis of relevant literature and interviews with stakeholders also showed that researchers not identifying and disclosing inventions is a government-wide challenge. For example, one stakeholder stated that researchers at federal labs generally have limited understanding of the patenting process, including an understanding of what constitutes

---

[47] The areas of identifying inventions, keeping track of inventions, and negotiating license agreements were selected for illustrative purposes only.

patentable subject matter and how to conduct a prior art search on the technology to determine whether it is patentable.[48]

DOD, DOE, NASA, and NIH officials stated that they are taking a variety of actions to help address this challenge. For example, some agency and lab officials stated that labs conduct training to educate researchers about the patenting process, inform researchers about requirements to disclose inventions, and incentivize them by acknowledging their efforts through awards and monetary incentives—such as potential royalty distributions— when their inventions reach commercial success.[49]

In addition, DOD, DOE, and NIH officials described their agencies' systems for keeping track of inventions developed in the labs as inadequate or in need of improvement. How agencies and labs keep track of such inventions can range from spreadsheets to sophisticated databases that manage all technology transfer activities, including keeping track of patented inventions and licenses. Currently, DOD has a decentralized approach to keeping track of inventions, which, according to DOD officials, needs improvement given how large the agency is.

Several stakeholders we interviewed also noted that the challenge of keeping track of inventions exists government-wide. According to some stakeholders, federal labs not only have inadequate systems to keep track of their own inventions but also limited information on the kinds of inventions being developed in federal labs across the government. The result is that agencies risk being unaware of research across the labs, which can limit their ability to leverage other federal research efforts. For example, one stakeholder stated that there can be research conducted independently at three or four labs under different agencies but little interaction among those labs about the research.

DOD, DOE, and NIH officials stated that they have made efforts to improve their current systems for keeping track of inventions. Specifically,

---

[48] Prior art is information relevant to a claimed invention and may include prior patents, patent applications, or nonpatent publications describing a technology, among other things.

[49] For example, the head of the agency or lab is required to pay each year the first $2,000 and thereafter at least 15 percent of the royalties or other payments to federal researchers from license agreements based on their inventions. 15 U.S.C. § 3710c.

DOE officials reported that they have developed a plan to leverage the capabilities of the iEdison reporting system to unify the agency's data management process. Air Force and NIH officials stated that they have contacted NASA, which has a centralized system for tracking inventions, about leveraging its expertise.[50] NASA officials reported that they have been hosting regular webinars with other agencies to determine whether NASA's tracking system could help meet other agencies' needs.

Furthermore, agency and lab officials and stakeholders noted that federal labs face challenges in negotiating license agreements because the licensing process is lengthy and uniquely regulated, which can deter companies from licensing federal inventions. Stakeholders stated that the federal licensing process can take anywhere from about 3 months to more than 2 years. Some stakeholders stated that from their point of view taking a year to negotiate a license agreement is too long. One stakeholder said that such lengthy processes are particularly difficult for start-ups, which often need to finalize license agreements in 3 months.

DOD, DOE, NASA, and NIH officials said they are taking steps to address companies' concerns about the time it takes to negotiate a license agreement. For instance, NASA, NIH, and Navy officials told us that they have developed model license agreements to help guide companies through the process, and NASA and NIH have special license agreements for start-ups to shorten the licensing process.[51]

For more detail on challenges in the seven areas of the patent licensing process that agency and lab officials and external stakeholders identified, see appendix II.

---

[50] NASA's system used in tracking patent licensing and other technology transfer activities is called the NASA Technology Transfer System. The system is operated at the lab level and is accessible to NASA officials at headquarters.

[51] The Startup NASA License and the agency's QuickLaunch provide standardized licensing terms and types of licenses that are available without negotiation. According to HHS officials, NIH also had a start-up license program with standardized non-negotiable terms from fiscal years 2012 to 2015, with the program ending after technology transfer activities at NIH were decentralized. However, HHS officials stated that the National Cancer Institute revamped the initiative and continues to offer it with the goal of reducing negotiation time periods.

Challenges in Prioritizing Patent Licensing

DOD, DOE, NASA, and NIH face challenges in prioritizing patent licensing as part of their agency missions. For example, DOD and DOE officials stated that an agency's mission affects patent licensing activities. DOD officials stated that the agency's primary mission is protecting the warfighter and that patent licensing is a secondary benefit to the agency. According to DOE officials, the nuclear security labs do not focus on patenting but instead on developing technologies associated with a weapons program.

In addition, several stakeholders we interviewed stated that some agencies and labs do not have a culture that prioritizes patent licensing. In particular, one stakeholder stated that at some federal labs, patent licensing is not reflected in performance evaluation management plans, which can help incentivize lab personnel to engage in patent licensing activities. A few stakeholders stated that at some labs where management does not prioritize patent licensing activities, researchers' careers can be negatively affected if they engage in patent licensing activities.

Some agency and lab officials stated that they have taken steps to overcome such challenges. For example, officials at one Navy lab stated that the lab has management support and nine patent attorneys to assist in the reviews of researchers' invention disclosures. Also, officials at one NIH lab stated that the lab has strong management support and a good royalty stream from successful inventions that pay for patenting and other reinvestments, which allows the lab to not draw from its appropriations.

NIST Has Reported Some Challenges Faced by Federal Labs in Areas of Patent Licensing but Has Not Fully Reported on the Range of Such Challenges

In its three most recent fiscal year summary reports to Congress, NIST identified some challenges faced by federal labs in areas of patent licensing and has assisted agencies in addressing challenges in their patent licensing activities. However, NIST does not fully report on the range of challenges that agency and lab officials and stakeholders identify.

NIST collaborates with agencies to gather patent licensing data for its summary reports to Congress. For example, according to agency officials, NIST engages with agencies to inform them about new requirements in technology transfer and helps them identify their successes in conducting technology transfer activities. NIST also provides administrative support to the FLC, which offers training to federal technology transfer specialists through workshops; publishes a desk reference on federal patent licensing, laws, and regulations; and has commissioned studies on efforts to develop federal inventions for commercial use.[52]

Further, NIST developed a survey in 2016 on agency technology transfer processes. NIST officials stated that the survey is aimed in part at improving federal labs' decisions on whether to spend money on applying for patents, whether patents will facilitate the commercialization of technology, and what data are needed to make those determinations. NIST officials stated that the agency continues to analyze the survey data and currently plans to report its findings in fiscal year 2018.

While NIST has identified in its annual summary reports to Congress some challenges that federal labs face in patent licensing and other technology transfer activities, it has not fully reported the range of challenges that agencies and labs face in patent licensing. For example, in its fiscal year 2015 summary report—its most recent report—on federal technology transfer, NIST reported that the federal intramural research

---

52 Federal Laboratory Consortium, *FLC Technology Transfer Desk Reference*; Federal Laboratory Consortium, *The Green Book: Federal Technology Transfer Legislation and Policy*, 5th ed. (Cherry Hill, N.J.: 2013); Indiana University, Kelley School of Business, Indiana Business Research Center, *Innovation Partnership Networks in the Midwest: A Pilot Project of the Federal Laboratory Consortium Midwest* (Bloomington, Ind.: Indiana University, 2013); and Raymond L. Sterling, *Utility Locating Technologies: A Summary Of Responses to a Statement of Need Distributed by the Federal Laboratory Consortium for Technology Transfer* (Federal Laboratory Consortium, 2000).

budget has been relatively consistent over the years but not that DOD, DOE, NASA, and NIH face challenges in prioritizing patent licensing as an agency mission.[53] The report also mentions that there is no uniform federal system for tracking research that employees in federal labs published but not that DOE, for example, has faced challenges in keeping track of inventions developed in its labs. In addition, we found that although the report mentions that the Department of Veterans Affairs is facing challenges with its labs disclosing inventions, it does not mention similar challenges at DOD. NIST officials stated that they were generally aware of the challenges identified by agency and lab officials and external stakeholders but had not considered including such challenges to a greater degree in the summary reports to Congress.

We have previously reported on Congress's goal to make the federal government more results oriented through reporting of agency performance information to aid decision making by agency executives, Congress, and program partners.[54] Specifically, we have reported how the effective implementation of good governance can help address government challenges in five key areas involving agency performance and management: (1) instituting a more coordinated and crosscutting approach to achieving meaningful results, (2) focusing on addressing weaknesses in major management functions, (3) ensuring that agency performance

---

[53] Department of Commerce, National Institute of Standards and Technology, *Federal Laboratory Technology Transfer Fiscal Year 2015: Summary Report to the President and the Congress* (Gaithersburg, Md.: December 2017). The federal intramural research budget includes the budget for intramural programs and the budget for contractor-operated labs.

[54] GAO, Managing for Results: Agencies Should More Fully Develop Priority Goals under the GPRA Modernization Act, GAO-13-174 (Washington, D.C.: Apr. 19, 2013); Managing for Results: GPRA Modernization Act Implementation Provides Important Opportunities to Address Government Challenges, GAO-11-617T (Washington, D.C.: May 10, 2011); Results-Oriented Government: GPRA Has Established a Solid Foundation for Achieving Greater Results, GAO-04-38 (Washington, D.C.: Mar. 10, 2004); and Agencies' Strategic Plans Under GPRA: Key Questions to Facilitate Congressional Review, GAO/GGD-10.1.16 (Washington, D.C.: May 1997).

information is useful and used in decision making, (4) sustaining leadership commitment and accountability for achieving results, and (5) engaging Congress in identifying management and performance issues to address.[55] By fully reporting the range of challenges in federal patent licensing—such as those outlined in this report—and including that information in its annual summary reports to Congress, NIST has the opportunity to further ensure that Congress is more aware of challenges that limit agencies' efforts in patent licensing and ways for potentially addressing those challenges. To identify these challenges, NIST could, for example, leverage its survey, past FLC studies, and agency reports.

FEDERAL AGENCIES AND LABS HAVE LIMITED INFORMATION ON PROCESSES, GOALS, AND COMPARABLE LICENSES TO GUIDE ESTABLISHING FINANCIAL TERMS IN PATENT LICENSES

Federal agencies and labs have limited information on processes, goals, and comparable licenses to guide establishing the financial terms in patent licenses. DOD, DOE, NASA, and NIH labs generally do not document their processes for establishing the financial terms of patent licenses and instead rely on the expertise of technology transfer staff. Furthermore, existing agency and lab guidance does not consistently link the practice of establishing license financial terms to the statutory goal of promoting commercial use of inventions. In addition, although many federal labs rely on comparable licenses to aid them in setting the terms of new licenses, labs have varying levels of access to information about such licenses.

---

[55] GAO-11-617T, and GAO, Government Performance: GPRA Modernization Act Provides Opportunities to Help Address Fiscal, Performance, and Management Challenges, GAO-11-466T (Washington, D.C.: Mar. 16, 2011).

Federal Agencies and Labs Have Limited Documentation of Their Processes for Establishing the Financial Terms of Patent Licenses

DOD, DOE, NASA, and NIH labs have limited documentation of their processes for establishing the financial terms of patent licenses. Such documentation is limited at both the agency level and the lab level.[56]

At the agency level, the four agencies we reviewed had some documentation on patent licensing in general, such as policies, procedures, guides, and handbooks, but had limited information on how to establish financial terms.[57] For example, the Air Force and the Navy had handbooks on technology transfer that include brief passages on financial terms. However, agency officials noted that these handbooks were either outdated or under revision.[58] At DOE, labs collaborated to develop two agency-level documents on patent licensing: one for lab officials on using equity in licenses and a licensing guide for licensees. These documents describe the general structure of various types of financial terms and, in the document on using equity, factors to consider regarding its use in a license, but do not discuss methods for establishing financial terms. NASA and NIH have policies and procedures for patent licensing that mention the types of financial terms that are normally found in licenses but do not cover other aspects, such as methods for establishing financial terms. All four agencies

---

[56] In addition, the *Technology Transfer Desk Reference*, made available to agencies and labs through the FLC, provides a list of factors to consider in setting financial terms but does not cover other aspects of the process, such as methods for establishing the financial terms.

[57] The agency-level documentation includes some that applies military department-wide (i.e., Air Force, Army, and Navy). GAO reviewed documents provided for coverage of four potential aspects of the process for establishing financial terms: (1) factors to consider, (2) sources of information, (3) methods for developing financial terms, and (4) use of specific types of financial terms. Factors may include a variety of considerations associated with the invention, the market, and the prospective licensee, such as the licensee's technical ability or the licensee's plan to address a public need. According to agency officials, license terms are typically negotiated based on specific circumstances in addition to any agency efforts to understand the value of the invention.

[58] The Navy handbook includes information on the types of financial terms that can be included and factors to consider but generally does not cover methods for establishing the terms, although it contained one example of guidance on how to set minimum annual fees.

reported that they gave their labs discretion to develop their own processes for establishing financial terms.

At the lab level, DOD, DOE, NASA, and NIH generally had not documented their processes for establishing financial terms in patent licenses. Based on documentation provided by NASA, NIH, and DOD, few labs at these agencies had issued additional documentation on the patent licensing process.[59] DOE labs had documented the patent licensing process in general, and 6 out of 17 DOE labs provided documentation that covered aspects of establishing financial terms. For example, one DOE lab document contained a set of licensing principles that help clarify what financial terms a license usually contains, their purpose, and how to structure the financial terms in patent licenses. In addition, agency and lab officials at NASA and DOE reported using tools, such as financial term calculators, at some of their labs, which aid technology transfer staff in valuing technologies.

Agency and lab officials reported that they generally rely on the expertise of technology transfer staff to establish and vet appropriate financial terms.[60] Accordingly, agencies and labs reported that they have taken some steps to develop, share, and retain expertise among staff in their technology transfer offices. The agencies we reviewed reported that some technology transfer staff participate in training opportunities provided by professional organizations like the Association of University Technology Managers (AUTM) or the Licensing Executives Society (LES), as well as the FLC and the agencies.[61] In addition, some agencies

---

59 DOD subcomponents and labs had some documentation of their process, including two handbooks, which had brief passages on financial terms, and some training materials, which in one case contained more detailed information on establishing financial terms.

60 Labs use a variety of methods to establish financial terms, ranging from identifying comparable licenses, or benchmarks, to direct valuation (e.g., calculations based on potential costs, sales, and other information). The financial terms often include up-front fees, minimum annual payments, royalties (often a percentage of sales), and milestone payments. Contractor-operated labs are also able to take equity in a licensee as part of the compensation for the license. For additional details on fiscal year 2014 licenses and their contents, see app. III.

61 AUTM is a nonprofit organization that supports and enhances the global academic technology transfer profession through education, professional development, partnering, and advocacy.

and labs reported that internal working groups and regular meetings are opportunities to share licensing expertise. At DOD, officials stated that on a case-by-case basis, labs may use the expertise of their partnership intermediary to help establish financial terms.

However, according to agency and lab officials and stakeholders, federal labs face challenges in acquiring, developing, and retaining expertise in patent licensing for their technology transfer offices. Specifically, some agency officials, lab officials, and stakeholders cited issues such as losing experienced technology transfer staff to retirement or to the private sector, having difficulties in hiring staff with expertise in part because of limited funding, and facing a limited pool of prospective employees to hire with the expertise to value and license inventions. A few stakeholders said that government training in the business aspects of patent licensing is inadequate and not widespread.[62] In addition, some stakeholders had concerns about consistency in licensing practices both within the labs and across labs. For example, some of these stakeholders said that the outcome of license negotiations can depend on the specific licensing professional handling the license. Varying levels of expertise may lead to inconsistency in licensing practices, including establishing financial terms, as can undocumented processes.

Under the federal standards for internal control, management should design control activities by, for example, clearly documenting them in management directives, administrative policies, or operating manuals, to achieve objectives and respond to risks.[63] Furthermore, documentation can

---

LES is an independent, professional organization that facilitates global intellectual property commerce through education, networking, standards development, and certification.

62 The extent of training varies by agency and lab. For example, NIH officials reported that the agency provided training on technology valuation agency-wide in 2017. On the other hand, a DOD lab official expressed concerns that the lab does not have expertise in establishing financial terms. Some lab officials also noted that they have to pay for their own membership to organizations such as AUTM, which offers on-demand training.

63 GAO, *Standards for Internal Control in the Federal Government*, GAO-14-704G (Washington, D.C.: September 2014).

act as a means to retain organizational knowledge and provide some assurance that an approach is operational across the lab or agency.[64]

Agency and lab officials stated that they had not documented their processes for establishing financial terms for various reasons. For example, lab officials stated that establishing financial terms is often complex and varies based on the specific circumstances applicable to each potential license, which may limit what can be documented. Some agency and lab officials stated that labs need flexibility in negotiating terms to make adjustments based on the circumstances and therefore officials do not want to be prescriptive. A few agency and lab officials also noted that there are benefits to having streamlined processes. Furthermore, a few agency and lab officials described negotiating license terms as a craft or art that requires expertise and said that documenting this will not enhance licensing by itself.

However, some agency and lab officials and stakeholders said that it is possible to document some aspects of the process. A few stakeholders we interviewed noted that even if each agreement is unique, it is still possible to develop guidelines or outline a methodology for establishing financial terms. A few agency and lab officials stated that they are investigating opportunities to standardize their processes or would be open to documenting them. For example, one agency official told us that the agency plans to update existing documents with specific information about royalty ranges so labs do not have to constantly "reinvent the wheel." Some labs also described steps that they take to establish financial terms, such as methods for valuing inventions, without being prescriptive. By documenting processes for establishing the financial terms of licenses while maintaining enough flexibility to tailor the specific terms of each license, the four agencies could have more reasonable assurance of consistency across their labs regardless of the expertise of staff.

64 The specific documentation needed will depend in part on the size, level of expertise, and hiring practices of each technology transfer office.

Federal Agency and Lab Documentation Does Not Consistently Link Financial Terms to the Goal of Promoting Commercial Use

Agency and lab documentation does not consistently link establishing financial terms in patent licenses to the goal of promoting commercial use of inventions. As noted above, federal law states that it is Congress's policy and objective to use the patent system to promote the commercialization and public availability of inventions, and that technology transfer, including federal patent licensing, is the responsibility of each laboratory science and engineering professional.[65]

Agency-level documentation at NASA contains a provision that clearly links establishing financial terms to the goal of promoting commercial use of inventions—that is, "terms should be negotiated that provide the licensee incentive to commercialize the invention." NIH's documentation mentions financial terms in the context of protecting the public from nonuse, which is one aspect of promoting commercial use, and also mentions the goal of obtaining a fair financial return on investment from the licensed invention.[66] DOD and DOE agency-level documents mention the general goal of promoting the commercial use of inventions without specifically linking it to the financial terms.[67] At the lab level, DOD documents generally do not address the goals for financial terms.[68] Of 17 DOE labs, 4 had a statement in their documentation to link financial terms to the goal of promoting commercial use of inventions.[69]

---

[65] See 15 U.S.C. § 3710; 35 U.S.C. § 200.

[66] According to NIH officials, specific terms, including financial terms, provide an incentive to develop inventions into marketable products that support public health. For example, NIH officials stated that minimum annual royalties incentivize a licensee to put a product on the market because such royalties are not immediately creditable and therefore remain a sunk cost for licensees until products are sold. In addition, NIH officials stated that their licensees are held accountable to detailed commercial development plans and timelines.

[67] While agency-level documents rarely discussed goals in the context of financial terms, two documents (one at each agency) had limited statements regarding goals for specific financial terms, such as equity, that are consistent with the goal of commercial use.

[68] One DOD lab training document stated that maximizing revenue is not the goal.

[69] Some other DOD and DOE labs had stated goals for license financial terms that did not directly connect to commercial use, such as making sure that the terms do not place an undue burden on the lab or provide sufficient benefit to the lab.

DOD, DOE, NASA, and NIH officials we interviewed stated that getting the technology into the marketplace is their primary goal in licensing but also mentioned other goals related to financial terms that support their mission. In addition, some agency and lab officials described using revenues from licenses as a means to provide a reward to inventors for their work or to obtain a fair return on investment for research conducted by federal agencies. Furthermore, lab officials we interviewed mentioned the flexibility of revenues from licenses as helpful in funding activities, such as additional research, training, and patent prosecution.

Some agency officials and stakeholders we interviewed expressed concerns about competing goals for establishing financial terms. For example, a few stakeholders stated that licensing professionals may be motivated to negotiate for increased license revenue because it reflects positively on them professionally. Further, some stakeholders expressed concerns about labs taking a short-term view of some licensees, particularly small companies, because they have less ability to pay initially and thus may offer less certain revenues.[70]

Our review of relevant economic literature and interviews with stakeholders suggest that license financial terms set with goals other than promoting commercial use in mind, such as short-term revenue maximization, may undermine that longer-term goal. For example, high up-front license fees typically provide more guaranteed short-term revenue to the licensor than other forms of payment but can also reduce the capital available to develop a product successfully. Labs with other goals in mind when establishing financial terms may be at risk of establishing them in ways that run counter to the goal of promoting commercial use.[71]

---

[70] On the other hand, a stakeholder and an agency official expressed concerns that labs may establish financial terms that are too low, which may provide little incentive to develop the technology.

[71] Some agency and lab officials stated that they may make adjustments to the financial terms based on the specific needs of the licensee. Some agency and lab officials reported that they take some steps to mitigate risks associated with competing goals, such as implementing conflict of interest procedures and appeals processes and preventing researchers from negotiating financials terms. We did not conduct a detailed assessment of the effectiveness of those measures in counteracting risks.

NIST plays an important role in providing regulations and guidance to agencies regarding patent licensing. Commerce has delegated to NIST the authority to promulgate implementing regulations pertaining to patenting and licensing at federal labs—that is, regulations that indicate how agencies are to implement statutory provisions, including the goal of, among other things, promoting commercial use of inventions. NIST has developed regulations, but they do not link the financial terms of federal patent licenses and the statutory goal of promoting commercial use of inventions.[72]

As the host of the FLC and a coordinator for the Interagency Working Group for Technology Transfer, NIST also plays a role in supporting the development of interagency guidance on patent licensing that covers, among other topics, establishing financial terms in licenses. However, existing interagency guidance provides limited information regarding the goals for financial terms. For example, the FLC desk reference contains a statement that links royalty rates to the goal of promoting commercial use but does not clarify how the goal applies to other financial terms. Furthermore, the FLC desk reference states that labs are entitled to market-based compensation for their intellectual property. However, licenses are structured differently to accomplish different goals and a primary focus on obtaining market-based compensation may undermine the goal of promoting commercial use.

As the lead agency on the government-wide effort to find commercial uses or practical applications for federally funded inventions, NIST has been delegated the responsibility to promulgate regulations pertaining to patenting and licensing at federal labs, including implementing the statutory goal of promoting commercial use. NIST officials stated that a change to the regulations could be made as part of an upcoming rule-making process. However, in doing so, a stakeholder and agency officials noted that any changes to the regulations should avoid prescriptive language that mandates specific practices. NIST officials also stated that they could update relevant guidance on this issue through one of their

---

[72] 37 C.F.R. pts. 401, 404.

current efforts. By clarifying the link between establishing federal patent license financial terms and the goal of encouraging commercial use, through the upcoming rule-making process and updating relevant guidance, NIST would have better assurance that financial terms in patent licenses are targeted to that goal.

Federal Agencies and Labs Have Varying Amounts of Information on Comparable Licenses, but Such Information Is Not Shared across Agencies

According to agency and lab officials, comparable license information can be used as a point of reference to guide establishing financial and other terms in new patent licenses. Just as real estate agents look at sales of comparable houses when setting the selling price of a house, patent licensing professionals can look at licenses for comparable inventions when determining what financial terms to include in a new license.

However, federal labs have varying amounts of information on comparable licenses when establishing financial terms. NASA and NIH each have an agency-wide system that enables each lab to access information from other labs at the agency, including the financial terms in previous licenses. NIH agency officials reported that technology transfer offices have access to thousands of previous licenses and refer to such information frequently to help establish the financial terms of new licenses. Labs at DOE and DOD are generally responsible for tracking their own licenses and do not have access to information on comparable licenses from other labs in their agencies. According to DOE officials, under DOE contracts and relevant law, license information at the agency's contractor-operated labs is considered business sensitive and a contractor-owned record that resides at the labs, which limits DOE's ability to share it. Officials at DOE and DOD's military departments reported that they have investigated and continue to investigate systems that would provide greater access to information on financial terms but have encountered some obstacles, such as network security requirements, that they have not yet overcome.

To bolster their access to comparable license information, some federal labs obtain private sector license information. For example, some lab officials we interviewed said that they have occasionally purchased benchmarking guides and access to other private sector license information through organizations such as AUTM and LES. According to some lab officials and stakeholders, private sector license information is useful for understanding acceptable royalty rates in industry and may cover certain technology areas or inventions that are new to the lab. However, access to private sector license information is typically ad hoc and can be limited by its cost, according to agency and lab officials. Some agency and lab officials stated that they would like increased access to private sector information on comparable licenses. For example, according to agency officials at DOE, there is an effort under way to obtain benchmark financial terms from labs and universities with comparable R&D portfolios.

Although lab officials and stakeholders said that private licensing information can be helpful for understanding financial terms acceptable to the market, using private license information may not always be appropriate for government licenses. Private licenses are often structured to maximize revenue for the licensor—not necessarily to promote commercial use or practical application, according to stakeholders. Our review of economic literature and interviews with stakeholders and agency officials suggest that licenses are structured differently to accomplish different goals. For example, a few stakeholders and agency officials noted that federal licenses would typically be less exclusive and have different financial terms than those in the private sector, where there is a greater emphasis on generating revenue from R&D investments.[73] Some stakeholders and agency officials also stated that in general the value of a government license may be different from that of a private license for a similar technology because of the rights the government retains on its

[73] One stakeholder noted that industry averages should be used as a guide to establishing royalty rates to avoid undercutting private licensing.

licenses. In addition, according to agency and lab officials and stakeholders, government inventions tend to be in an earlier stage of development than those in the private sector, potentially making it more difficult to find licenses for comparable inventions in the private sector.

Some agency and lab officials and a few stakeholders stated that it would be valuable for federal labs to have greater access to information on financial terms in government licenses to help establish a benchmark for financial terms. Our analysis of approximately 21,000 patents assigned to DOD, DOE, NASA, and NIH and issued since 2000 shows that different agencies may patent inventions in similar technology fields.[74] All four agencies we reviewed had patented inventions in 26 of 35 technology fields covered by the patents, and all had 10 or more patents in 9 of the 35 technology fields.[75] DOD and DOE, including DOE contractor-operated labs, had more patents in a wider range of fields than the other agencies. On the other hand, HHS's patents are more focused on fields such as biotechnology and medical technology. However, even in the area of biotechnology, there were hundreds of patents issued to the other three agencies. Although other information would be needed to determine whether the agencies' inventions are truly comparable, their having patents in the same technology fields suggests that some government-wide information on financial terms could be useful to federal labs.

Under internal control standards for the federal government, management should externally communicate the necessary quality information to achieve the entity's objectives; this includes communicating with and obtaining quality information from external parties using established reporting lines.[76] The four agencies we reviewed communicate and share information through several collaborative efforts to improve

---

[74] NIH patents are assigned to HHS. Accordingly, all HHS patents were used for determining the extent to which NIH is patenting in specific technology areas. DOE and NASA patents include patents assigned to contractor-operated labs. Design and plant patents were excluded from the analysis of WIPO technology fields.

[75] In 22 of these 26 technology fields, we identified more than 200 total patents. Substantial shares of these patents were assigned to multiple agencies, not concentrated at a single agency.

[76] GAO-14-704G.

federal patent licensing, including the FLC and the Interagency Working Group for Technology Transfer. For example, agency officials said they share experiences, ideas, and best practices related to patent licensing informally through these groups. However, there is no formal sharing of information on financial terms in patent licenses among federal labs, according to NIST officials.

We have previously reported that federal agencies engaged in interagency collaborative efforts should identify and address needs by leveraging their resources to obtain additional benefits that would not be available if they were working separately.[77] NIST plays a leading role in these interagency collaborative efforts on patent licensing, including gathering and sharing information among the labs. As the administrative host for the FLC, NIST has already supported an effort to share information about available technology. NIST is also responsible for gathering information from technology transfer agencies, including gross license income, and submitting summary reports to Congress annually and sharing them with the public.[78] Furthermore, NIST has initiated a survey of practices at federal technology transfer offices and shared some preliminary information with the agencies. By facilitating the formal sharing of comparable license information, NIST could help provide agencies and labs with benchmarks for evaluating which financial terms are best suited to licensing inventions successfully.

NIST officials stated that gathering and sharing comparable license information could be done as part of their existing efforts but that there are obstacles to doing so. Specifically, NIST officials stated that this effort would add to the reporting burdens of agencies, may require additional resources, and would need to take into account data security and proprietary information considerations. Agency officials also stressed that any effort to share license terms would have to ensure that confidential and

---

[77] GAO, Results-Oriented Government: Practices That Can Help Enhance and Sustain Collaboration among Federal Agencies, GAO-06-15 (Washington, D.C.: Oct. 21, 2005).

[78] License income is the amount of income that agencies receive as a result of the financial terms in their licenses. Currently, NIST does not gather granular information on the individual financial terms.

proprietary information from licensees, including specific financial terms from a particular license, is not divulged.[79]

CONCLUSION

Federal labs under DOD, DOE, NASA, and NIH face challenges at various stages of the patent licensing process, and agencies have taken some steps to address such challenges. For example, ensuring that researchers identify and disclose inventions is a government-wide challenge, according to interviews with external stakeholders and our analysis of relevant literature. However, such challenges in federal patent licensing are not fully reported by NIST, the lead agency delegated by Commerce to provide annual summary reports to Congress on federal technology transfer activities. By fully reporting the range of these challenges that agencies and labs face, NIST can ensure that Congress has greater awareness of these challenges. To help identify these challenges, NIST could, for example, leverage its survey of practices at federal technology transfer offices, past FLC studies, and agency reports.

In addition, DOE, DOD, NASA, and NIH documentation does not consistently link establishing financial terms in patent licenses to the statutory goal of promoting commercial use. As the lead agency on the government-wide effort to find commercial uses or practical applications for federally funded inventions, NIST has been delegated the responsibility to promulgate regulations pertaining to patenting and licensing at federal labs, including implementing the statutory goal of promoting commercial use. By clarifying the link between establishing patent license financial terms and the goal of encouraging commercial use, through the upcoming rule-making process and updating relevant guidance, NIST would have better assurance that financial terms in patent licenses are targeted to that goal.

---

79 Access to such information would have to balance the need to keep proprietary business information confidential as required by the Federal Trade Secrets Act. See 18 U.S.C. § 1905.

Further, federal labs have varying amounts of information on comparable government licenses when establishing financial terms. However, there is no formal sharing of information on financial terms in patent licenses among federal labs, according to NIST officials. NIST plays a leading role in interagency collaborative efforts on patent licensing, including gathering and sharing information among the labs. By facilitating the formal sharing of comparable license information, NIST could help provide agencies and labs with benchmarks for evaluating which financial terms are best suited to successfully licensing inventions.

To establish financial terms, DOD, DOE, NASA, and NIH labs rely on the expertise of their technology transfer staff and take a number of steps to build and share expertise, but had limited documentation of their processes for establishing the financial terms of patent licenses. Agency and lab officials explained that there is a need for flexibility, and thus not every aspect of their processes can be documented in detail. By documenting processes for establishing the financial terms of licenses while maintaining enough flexibility to tailor the specific terms of each license, the four agencies could have more reasonable assurance of consistency across their labs regardless of the expertise of staff.

Recommendations for Executive Action

We are making seven recommendations, including three to Commerce and one each to DOD, DOE, NASA, and NIH:

- The Secretary of Commerce should instruct NIST to fully report the range of challenges in federal patent licensing, such as those outlined in this report, by, for example, leveraging its survey of practices at federal technology transfer offices, past FLC studies, and agency reports and including that information in its summary reports to Congress. (Recommendation 1)
- The Secretary of Commerce should instruct NIST to clarify the link between establishing patent license financial terms and the

goal of promoting commercial use, through appropriate means, such as the upcoming rule-making process and updating relevant guidance. (Recommendation 2)

- The Secretary of Commerce should instruct NIST to facilitate formal information sharing among the agencies to provide federal labs with information on financial terms in comparable patent licenses, as appropriate. (Recommendation 3)
- The Secretary of Defense should ensure that the agency or its labs document processes for establishing license financial terms, while maintaining flexibility to tailor the specific financial terms of each license. (Recommendation 4)
- The Secretary of Energy should ensure that the agency or its labs document processes for establishing license financial terms, while maintaining flexibility to tailor the specific financial terms of each license. (Recommendation 5)
- The Administrator of NASA should ensure that the agency or its labs document processes for establishing license financial terms, while maintaining flexibility to tailor the specific financial terms of each license. (Recommendation 6)
- The Director of NIH should ensure that the agency or its labs document processes for establishing license financial terms, while maintaining flexibility to tailor the specific financial terms of each license. (Recommendation 7)

AGENCY COMMENTS

We provided a draft of this report to Commerce, DOD, DOE, NASA, and NIH for review and comment. All provided written responses, which are reproduced in appendixes IV-VIII. Commerce and NIH also provided technical comments, which we incorporated as appropriate.

Commerce agreed with all three of our recommendations to the agency. In general, the agency stated that it will work through interagency groups, such as the Interagency Working Group for Technology Transfer and the FLC, to address our recommendations, including by creating a specific section in its annual reports to Congress with more details on challenges agencies and labs face in patent licensing and by examining and implementing solutions to facilitate the sharing of information among agencies. According to Commerce, such solutions could include identifying licensing officers who have expertise and creating a community of practice in which they can share best practices and approaches for establishing license terms.

DOD, DOE, and HHS agreed, and NASA partially agreed, with the recommendation that they or their labs document processes for establishing financial terms in patent licenses. In its written response, DOD said it will direct the military departments and appropriate defense agencies to have their labs establish documentation of their licensing processes as appropriate. In their written comments, DOE, HHS, and NASA noted the complexity and nuances associated with negotiating license agreements, such as understanding the market for the technology and the level of risk involved. Further, DOE and NASA noted challenges that limit their ability to document processes and emphasized the importance of maintaining flexibility in establishing financial terms in patent licenses. We agree that some flexibility in establishing financial terms of patent licenses is important. DOE, HHS, and NASA all identified steps they would take to ensure that at least some processes for establishing financial terms are documented.

We are sending copies of this report to the appropriate congressional committees; the Secretaries of Commerce, Defense, and Energy; the Administrator of NASA; and the Director of NIH. In addition, the report is available at no charge on the GAO website at http://www.gao.gov.

If you or your staff have any questions about this report, please contact me at (202) 512-3841 or neumannj@gao.gov. Contact points for our Offices of Congressional Relations and Public Affairs may be found on the last page of this report. GAO staff who made key contributions to this report are listed in appendix IX.

Sincerely yours,

John Neumann
Director, Natural Resources and Environment

APPENDIX I: EXAMPLES OF INVENTIONS DEVELOPED IN FEDERAL LABS

Source: GAO summary of agency information, National Institutes of health, courtesy of the Australian Institute of marine Sciences (photo, upper left); Department of Defense (photo, upper right); NASA (photo, ower left); DOE (photo, lower right).| GAO-8-327.

Figure 3. Examples of Inventions Developed in the Department of Defense, Department of Energy, National Aeronautics and Space Administration, and National Institutes of Health Federal Laboratories.

Figure 3 presents examples of inventions developed in federal laboratories under the Department of Defense, Department of Energy, National Aeronautics and Space Administration, and National Institutes of Health.

APPENDIX II: SELECTED DESCRIPTIONS OF CHALLENGES FEDERAL LABS FACE IN PATENT LICENSING

The following are additional descriptions of challenges in the seven areas of the patent licensing process as well as challenges in prioritizing patent licensing faced by federal laboratories (lab) that were identified by external stakeholders and by agency and lab officials at the Department of Defense (DOD), Department of Energy (DOE), National Aeronautics and Space Administration (NASA), and the National Institutes of Health (NIH)—as well as steps agencies and labs have taken to address those challenges.

Challenges in the Seven Areas of the Patent Licensing Process

Identifying Inventions

DOD, DOE, NASA, and NIH officials reported challenges in identifying inventions that lab researchers developed. When a federal researcher does not disclose to lab officials an invention developed in a federal lab, the opportunity to assess the invention's potential for commercial use may be lost.

Federal officials cited various reasons why researchers do not disclose inventions. For instance, several DOD, DOE, NASA, and NIH agency and lab officials stated that some researchers do not have adequate training in identifying potentially patentable inventions. Some agency and lab officials pointed to other reasons why invention disclosures may not be filed, such as researchers not having enough incentive to disclose their inventions. Navy officials stated that researchers are often intimidated by the overall invention disclosure process and tend to focus on their research rather than consider what could be patentable. Officials at one NASA lab noted that they have come across a few contractor employees who do not see the benefit of filing invention disclosures, and sometimes researchers are too busy to engage in the patenting process. According to National Institute of Standards and Technology (NIST) officials, some researchers decide not to disclose an invention because they believe filing a patent application, which includes a filing fee, could take away money from the research itself, and most federal researchers are not motivated by the potential for receiving royalty distributions.[1]

Our analysis of relevant literature and interviews with stakeholders also showed that researchers not identifying and disclosing inventions is a government-wide challenge. One stakeholder stated that researchers at federal labs generally have limited understanding of the patenting process, including an understanding of what constitutes patentable subject matter and how to conduct prior research on the technology to determine whether it is patentable.[2]

DOD, DOE, NASA, and NIH agency and lab officials stated that they are taking a variety of actions to help address these challenges. For example, some agency and lab officials stated that labs conduct training to educate researchers about the patenting process, inform researchers about statutory requirements to disclose inventions, and incentivize them by acknowledging their efforts through awards and monetary incentives when their inventions reach commercial success.

---

[1] For example, the head of the agency or lab is required to pay each year the first $2,000 and thereafter at least 15 percent of the royalties or other payments to federal researchers from license agreements based on their inventions. 15 U.S.C. § 3710c.

[2] Prior art is information relevant to a claimed invention and may include prior patents, patent applications, or nonpatent publications describing a technology, among other things.

Keeping Track of Inventions

DOD, DOE, and NIH officials described their agencies' systems for keeping track of inventions developed in the labs as inadequate or in need of improvement. How agencies and labs keep track of such inventions can range from spreadsheets to sophisticated databases that manage all technology transfer activities, including keeping track of patented inventions and licenses.

Currently, DOD has a decentralized approach to keeping track of inventions, which, according to DOD officials, needs improvement given how large the agency is. Each military department has its own systems to track and store information on inventions developed in the labs.[3] Officials from DOD and the departments describe the systems as inadequate to keep track of the agency's inventions. For example, Navy officials described the department's in-house system to track inventions as "plagued by outages" and thus ineffective. According to officials, the Army funds systems that track inventions, but these systems are different from each other and not connected to headquarters and have been suspended since 2015.[4]

We have previously reported on federal agencies' challenges in monitoring technology transfer activities, including tracking inventions developed in the federal labs.[5] Several stakeholders we interviewed also noted that keeping track of inventions is a government-wide challenge. According to some stakeholders, federal labs not only have inadequate systems to keep track of their own inventions but also limited information

---

[3] DOD and its departments use various systems to keep track of inventions developed in their labs, including internally developed Access-based systems, low-budget commercial software systems, and Excel spreadsheets.

[4] Army officials reported that the department's patent and technology transfer database was shut down by Army network security officials in 2015 because the system did not meet Army network security requirements. According to Army officials, the database did not track other key information—such as patent applications, disclosures, and licenses. While the Army labs have generally used spreadsheets to track their inventions, officials stated that they do not know whether all of the labs have put the information into spreadsheets and how far back the data have been kept.

[5] GAO, Intellectual Property: Federal Agency Efforts in Transferring and Reporting New Technology, GAO-03-47 (Washington, D.C.: Oct. 31, 2002), and Federal Research: DOE Is Addressing Invention Disclosure and Other Challenges but Needs a Plan to Guide Data Management Improvements, GAO-15-212 (Washington, D.C.: Jan. 30, 2015).

on the kinds of inventions being developed in federal labs across the government. The result is that agencies risk being unaware of research across the labs, which can limit their ability to leverage other federal research efforts. One stakeholder specifically noted that the Interagency Edison (iEdison) reporting system—which allows federal grantees and contractors to report federally funded inventions to the agency that issued the funding award, including inventions developed by some contractor-operated labs—is difficult to navigate and needs improvement. Another stakeholder stated that there can be independent research at three or four labs under different agencies but little interaction among those labs about the research. Information on federal lab inventions can also be accessed publically through the Federal Laboratory Consortium (FLC) website; however, NIST officials stated that the website's information on inventions relies on agencies to submit accurate information, which may be limited by the agencies' tracking systems.[6]

DOD, DOE, and NIH officials stated that they have made efforts to improve their current systems. For example, since our 2015 report on the agency's challenges with its data management systems that track federally funded inventions, DOE officials reported that they have developed a plan to leverage the capabilities of the iEdison reporting system to unify the agency's data management process.[7] While DOD officials stated that the agency has been unsuccessful in purchasing software to track inventions across the agency, Air Force officials said they are developing a pilot program and seeking new software to manage the Air Force's inventions, and they expect the pilot program to increase the number of invention disclosures. Air Force and NIH officials stated that they have contacted NASA, which has a centralized system for tracking inventions, about leveraging the agency's expertise.[8] NASA officials reported that they have

---

[6] The Federal Laboratory Consortium's website, FLCBusiness.com, allows external users to search for technologies available for potential licensing throughout the federal government.

[7] See GAO-15-212.

[8] NASA's system used in tracking patent licensing and other technology transfer activities is called the NASA Technology Transfer System. The system is operated at the lab level and is accessible to NASA officials at headquarters.

been hosting regular webinars with other agencies to determine whether NASA's tracking system could help meet other agencies' needs.

Selecting Inventions to Patent

DOD, DOE, NASA, and NIH agency and lab officials cited selecting inventions to patent as a challenge because of the expense of patenting fees.[9] According to some agency and lab officials we interviewed, fees paid to the United States Patent and Trademark Office (USPTO) affect their decision on whether to patent an invention. For example, DOE officials stated that budget constraints force them to make decisions about whether they should file a patent or engage in other agency activities. NIH officials stated that the agency maintains fewer patents because of the patent maintenance fees and the agency's tight budgets.[10]

NASA officials reported that one step the agency is taking to deal with the costs of maintaining its issued patents is to identify technologies with low licensing potential and allow the patents to expire if they fail to attract licensees. NASA has created a searchable database that catalogs thousands of expired NASA patents already in the public domain, making them freely available to industry for unrestricted commercial use.

---

[9] The United States Patent and Trademark Office charges federal agencies all patent fees at the same rate as large corporations, including filing, examination and maintenance fees. The first maintenance fee is due at 3.5 years at a cost of $1,600, the second maintenance fee is due at 7.5 years at a cost of $3,600, and the third maintenance fee is due at 11.5 years at a cost of $7,400. Under the patent regulations, government organizations are classified as large entities and therefore pay undiscounted maintenance fees. Most fees, including maintenance fees, are discounted by 50 percent for small entities and by 75 percent for micro entities. U.S. institutions of higher education generally qualify for discounted fees. The agency sets and adjusts fees set by Congress in 35 U.S.C. § 41. Specifically, 35 U.S.C. § 41(h)(1) authorizes discounted fees only for small business concerns, independent inventors, and nonprofit organizations.

[10] In addition, stated NIH officials, high patent fees can affect research collaborations with universities and federal labs because USPTO also assesses undiscounted patent fees for such collaborations, which can become a burden for universities that would normally pay lower patent fees by filing individually. According to USPTO officials, offering discount eligibility to federal agencies may raise legal concerns with respect to international obligations.

Attracting Potential Licensees

Federal labs under DOD, DOE, NASA, and NIH face challenges that limit their ability to attract potential licensees, according to agency and lab officials. Even officials at NASA, described by NIST officials as one of the best agencies in promoting its inventions to industry, said the agency is not selecting among multiple licensees and would like to have more companies license its patents.

There are various reasons why federal labs struggle to attract companies interested in licensing their inventions, according to agency and lab officials we interviewed. First, several agency and lab officials cited that the number of entities that want to license inventions is generally not large. Second, some agency and lab officials identified inadequate promotion of federal inventions and licensing opportunities to companies, including start-ups, as a factor. Third, some agency and lab officials also noted that their inventions are often in the early stages of development and thus pose more of a risk for companies to license.

Based on our analysis of relevant literature and interviews with stakeholders, difficulty in attracting industry to license inventions developed in federal labs is a government-wide challenge. According to several stakeholders, industry perceives federal labs as not friendly to the private sector when it comes to patent licensing, especially for start-ups. For example, one stakeholder said that it is rare that federal agencies want to license to a start-up, and that more often the labs want a "safer route" by licensing inventions to large companies that already have a steady revenue stream. Another stakeholder said that DOE's contractor-operated labs in particular tend to not issue exclusive licenses to start-ups and prefer to license to large companies because the agency sees those companies as presenting less of a risk.[11] In addition, stakeholders stated that federal inventions are often not yet commercially viable, which can deter companies from licensing federal inventions. One stakeholder, for

[11] In general, according to stakeholders, exclusive licenses help protect start-ups as they try to develop the technology, whereas large companies that have more resources are generally more receptive to nonexclusive licenses. Nonexclusive licenses give companies more protection from patent infringement because more than one licensee can be granted rights to the invention.

example, stated that NASA officials may think that NASA technology is more developed than it is and therefore underestimate how long it will take a company to develop it for practical application, the millions of dollars needed to develop it, and whether it can be manufactured for commercial use.

DOD, DOE, NASA, and NIH officials stated that they are taking steps to attract potential licensees by, for example, conducting local outreach to attract companies and working on improving their databases so that companies can learn about federal inventions available for licensing. For instance, NASA officials stated that the agency's comprehensive database accessible to potential licensees uses a wide variety of search criteria and attracted 6 million unique visitors in 2016.

Negotiating the License Agreement

Agency and lab officials and stakeholders noted that federal labs face challenges in negotiating the license agreement because the process is (1) lengthy and (2) uniquely regulated, which can deter companies from licensing federal inventions.

Patent Licensing Process Timelines

Stakeholders stated that the federal licensing process can take anywhere from about 3 months to more than 2 years. Some stakeholders stated that from their point of view taking a year to negotiate a license agreement is too long. One stakeholder said that such lengthy processes are particularly difficult for start-ups, which often need to finalize license agreements in 3 months. Another stakeholder noted that the federal government in general does not understand how urgent it is for companies to complete the licensing process in a timely manner. Although actions on the part of both the labs and companies can cause delays, if the overall process is time-consuming, prospective licensees will tend to move onto something else instead, according to agency and lab officials and stakeholders.

Based on our analysis of licensing information provided by the agencies, we found that the amount of time from receipt of an application for a license to signature of the license by the lab varies widely. Specifically, based on this measure of the length of the process, approximately 60 percent of 132 licenses effective in fiscal year 2014 took at most 6 months for DOD, DOE, NASA, and NIH labs to process.[12] Officials at one Navy lab stated that issuing an invention license to a company within 6 months is "highly unusual," and officials at one NASA lab stated that the fastest they have issued a license was a week because the start-up was prepared and ready to go. For more on our analysis of licensing information from DOD, DOE, NASA, and NIH, see appendix III.

Regulated Patent Licensing Process

Several agency and lab officials also noted that federal regulations associated with patent licensing can deter companies from licensing federal inventions. Such regulations include requirements that are unique to federally funded and federally owned inventions, including that products arising from the invention must be substantially manufactured in the United States and that the government may retain rights to the invention and terminate the license agreement if the licensee does not take steps to commercialize the technology. In particular, NASA officials stated that venture capital firms sometimes oppose the government retaining rights for federal technology used by start-ups that they fund. According to DOD and DOE officials, federal regulations require a level of documentation or explanation that can deter some companies from licensing inventions developed in federal labs. Based on interviews with stakeholders, as well as our analysis of relevant literature, company concerns about federal regulations is a government-wide challenge that federal labs face in licensing their inventions.

For example, according to NIST officials, the U.S. manufacturing requirement can influence whether companies consider licensing federal inventions, because manufacturing in the United States can be more

---

[12] The number of licenses at DOD, DOE, NASA, and NIH that we reviewed only include commercial licenses, which give the licensee the right to sell a product incorporating a licensed invention.

expensive than manufacturing in other countries.[13] NIST officials also stated that some prospective licensees initially become concerned when they are told about march-in authority, because it applies to federally funded inventions and contractors. However, once companies are told that it is a legal requirement and that the provision has never been exercised, they generally become more comfortable with it.[14]

DOD, DOE, NASA, and NIH agency officials said they are taking steps to address companies' concerns about the time it takes to negotiate a license agreement and their unfamiliarity with federal licensing requirements. For instance, NASA, NIH, and Navy officials told us they have developed model license agreements to help guide companies through the process, and NASA and NIH have special license agreements for start-ups to shorten the licensing process.[15] Also, DOE created an agency-wide licensing guide to help prospective licensees navigate federal licensing requirements.[16]

Monitoring Licensee Performance

DOD, DOE, NASA, and NIH agency and lab officials we interviewed identified limited resources and inadequate monitoring systems as factors that make it difficult to monitor licensee performance.

---

13 According to NIH officials, the U.S. manufacturing requirement can affect companies that only have established manufacturing facilities located outside the United States.

14 According to a few agency and lab officials we interviewed, the threat of march-in authority, which may only be used in circumstances specified by statute, has been helpful in making sure contractors develop inventions for commercial use. We have previously reported on federal implementation of march-in authority: GAO, *Federal Research: Information on the Government's Right to Assert Ownership Control over Federally Funded Inventions*, GAO-09-742 (Washington, D.C.: July 27, 2009).

15 The Startup NASA License and the agency's QuickLaunch provide standardized licensing terms and types of licenses that are available without negotiation. According to HHS officials, NIH also had a start-up license program with standardized non-negotiable terms from fiscal years 2012 to 2015, with the program ending after technology transfer activities at NIH were decentralized. However, HHS officials stated that the National Cancer Institute revamped the initiative and continues to offer it with the goal of reducing negotiation time periods.

16 Department of Energy, Technology Transfer Working Group, *Licensing Guide and Sample License*, (published by Innovation: America's Journal of Technology Commercialization and Technology Ventures Corporation, no date).

NASA and NIH officials reported that the number of license agreements has increased in their labs and that they do not have enough resources to monitor licenses.[17] DOD officials stated that the agency's technology transfer offices have traditionally been understaffed and that the agency's monitoring systems are inadequate for tracking the status of issued licenses. Officials at one DOE lab stated that collecting royalties from licensees can be difficult because the lab does not have enough funds to support that activity. In addition, agencies may rely on the same systems they use to keep track of inventions to monitor licensee performance, and as previously discussed, these systems are in need of improvement.

Some stakeholders we interviewed noted that monitoring licensee performance is a government-wide challenge. They explained that sometimes licensees do not pay fees if they are not contacted, and a few stakeholders stated that federal labs have limited funding and resources to monitor contracts effectively. One stakeholder recalled one agency that did not communicate with a licensee for 2 years after the license agreement was signed. According to another stakeholder, ineffective monitoring of licensee performance may limit federal labs' ability to determine whether a company is developing federal inventions for commercial use per the terms and conditions of the license agreement.

Some agency and lab officials stated that they have taken steps to regularly monitor licensees. In particular, at NASA and NIH—where monitoring of licensee performance is centralized at the agency level—officials have programed systems to remind staff to check on licensee performance.

Measuring Licensing Outcomes

Federal labs, including those under DOD, DOE, NASA, and NIH, also face challenges in effectively measuring patent licensing outcomes, based on our interviews with stakeholders and analysis of relevant literature.

---

[17] NASA officials stated that the agency's centralized automated system produces reports on labs' monitoring efforts for the agency to review. NIH officials stated that establishing a centralized office dedicated to monitoring licensee performance has helped them assess whether licensees are developing NIH inventions.

According to one stakeholder, labs need metrics to assess whether a licensee has made progress on developing the invention for commercial use and whether the lab needs to get the license back and give it to another company.

However, some stakeholders we interviewed stated that although the 2011 presidential memorandum on technology transfer called for strategies to establish metrics, federal labs are still struggling to implement metrics for measuring technology transfer outcomes, including patent licensing activities. Stakeholders we interviewed and our analysis of relevant literature have indicated that federal labs in general track the numbers of patents, licenses, and revenues instead of using metrics that identify direct economic impacts from patent licensing and other technology transfer activities. In agencies where such metrics do exist, they may be applied inconsistently across labs. For example, officials at one DOE lab stated that DOE metrics are generally not consistent across the agency's labs.[18]

DOD, DOE, NASA, and NIH agency officials stated that they are working to improve their metrics and incorporate metrics beyond tracking numbers of patents, licenses, and revenues. For example, in addition to measuring the numbers of patents and licenses issued, NASA and Air Force officials stated that they are also measuring factors that affect the length of time it takes for their labs to process licenses. Such information, officials said, will help them expedite the licensing process.

Prioritizing Patent Licensing as an Agency Mission

DOD, DOE, NASA, and NIH face challenges in prioritizing patent licensing as part of their agency missions, which can affect the entire patent licensing process.

---

[18] DOE officials stated that the agency's technology transfer office has a metrics committee that is working to streamline the process of gathering information from its labs to help DOE measure its patent licensing outcomes. Also, DOE's Advanced Research Projects Agency—Energy collects data on the formation of new companies and funding from the private sector in measuring outcomes of inventions reaching the marketplace.

For example, DOD and DOE agency and lab officials stated that an agency's mission affects patent licensing activities. DOD officials stated that the agency's primary mission is protecting the warfighter and that patent licensing is a secondary benefit to the agency. According to DOE officials, the nuclear security labs do not focus on patenting but instead on developing technologies associated with a weapons program.

In addition, several stakeholders we interviewed stated that some agencies and labs do not have a culture that prioritizes patent licensing. In particular, one stakeholder stated that at some federal labs, patent licensing is not reflected in performance evaluation management plans, which can help incentivize lab personnel to engage in patent licensing activities. A few stakeholders stated that at some labs where management does not prioritize patent licensing activities, researchers' careers can be negatively affected if they engage in patent licensing activities.

DOD, DOE, NASA, and NIH agency and lab officials cited limited resources to conduct the range of activities related to patent licensing. For example, sometimes there is just one person at a DOD lab overseeing technology transfer activities, according to DOD agency and lab officials. Officials at one NIH lab stated that many labs across the agency do not receive enough royalties to offset their patent licensing costs. In its fiscal year 2015 report—its most recent report—to Congress on federal technology transfer activities, NIST reported that the federal intramural research budget, which include patent licensing activities, has generally not increased in the past 4 fiscal years.[19] Several agency and lab officials stated that budget constraints affect the extent to which they can engage in patent licensing activities—including patent enforcement, which can cost millions of dollars and presents challenges for federal labs, according to DOE officials.

Some agency and lab officials stated they have taken steps to overcome such challenges. For example, officials at one Navy lab stated that the lab has management support and nine patent attorneys to assist in the reviews of researchers' invention disclosures. Also, officials at one

---

[19] The federal intramural research budget includes the budget for intramural programs and the budget for contractor-operated labs.

NIH lab stated that the lab has strong management support and a good royalty stream from successful inventions that pay for patenting and other reinvestments, which allows the lab to not draw from its appropriations.

APPENDIX III: PATENT LICENSE SUMMARY FOR LICENSES EFFECTIVE IN FISCAL YEAR 2014

Tables 1 through 3 and figures 4 through 6 are based on 222 patent licenses that became effective in fiscal year 2014, and associated data, provided by the Department of Defense (specifically the Army, Navy, and Air Force), Department of Energy, National Aeronautics and Space Administration, and National Institutes of Health.[1] They include both data provided by the agencies[2] and information compiled directly from the licenses.[3] The tables and figures are provided for informational purposes and are not generalizable to all patent licenses.

---

[1] The tables and figures in this appendix exclude amendments to patent licenses, joint ownership agreements, assignment agreements, and government use licenses. Although we sought to identify all patents licenses that became effective in fiscal year 2014, we were unable to examine the underlying systems containing the data to confirm that no licenses were missing. From these data, we identified 222 licenses that became effective in fiscal year 2014 and requested signed copies. In some cases, federal laboratories (lab) were unable to provide all of the information requested either because it is not regularly tracked in their systems or was not available for specific licenses. To assess the reliability of the data associated with these licenses, we asked agency and lab officials general questions about the accuracy and completeness of the data and requested that agencies confirm specific information about licenses.

[2] These include the exclusivity of the license, the date on which the lab received an application for the license, the size of the licensees (i.e., whether they qualified as small businesses or start-ups), and patents issued. We merged the list of issued patents provided by the agencies with data from the PatentsView database to obtain information on the associated World Intellectual Property Organization technology fields.

[3] Using the signed licenses, we confirmed information, such as the effective date and the type of license, and compiled information about the financial terms. For 132 of these licenses that were commercial (i.e., they allowed for the sale of products incorporating the licensed invention) and for which data on the application date was available, we compared the application date to the date the lab signed the license as a proxy for the length of the process. Some contractor-operated labs do not use a formal application and were thus unable to provide such dates or provide the dates when they received commercialization plans or when terms were first discussed. Seven licenses were excluded based on data quality issues.

Table 1. License Types by Agency, Exclusivity, and Size of Licensee for Licenses Effective in Fiscal Year 2014

		Agency			Exclusive[a]			Licensee size[b]			
License type	DOD[e]	DOE	NASA	NIH	Yes	No	Small business	No data[c]	Start-up	No data[c]	*Total*
Commercial[d]	33	66	19	32	78	72	107	9	30	55	*150*
Noncommercial[d]	0	50	13	9	12	60	54	0	4	19	*72*
Total	33	116	32	41	90	132	161	9	34	74	222

Legend: DOD = Department of Defense; DOE = Department of Energy; NASA = National Aeronautics and Space Administration; NIH = National Institutes of Health.

Source: GAO analysis of DOD, DOE, NASA, and NIH licenses. | GAO-18-327.

[a] This category identifies whether licenses are exclusive in full or in part.

[b] A small business is defined under 37 C.F.R. 404.3(c). In this case, a start-up is defined as "a new company formed specifically to license and develop the technology being licensed."

[c] For some licenses, data on licensee size were not tracked or provided by the agency or lab.

[d] Commercial licenses give the licensee the right to develop and sell a commercial product incorporating a licensed invention. Noncommercial licenses include research, evaluation, and option licenses, which do not allow for the sale of a commercial product without obtaining another license.

[e] DOD licenses include only those identified from data provided by the Army, Navy, and Air Force for their respective labs.

Table 2. Top 10 Technology Fields for Issued and Licensed Patents in Fiscal Year 2014 Licenses at DOD, DOE, NASA, and NIH

Primary WIPO field	Number of issued patents that were licensed[a]	Number of licenses associated with those patents
Biotechnology	93	49
Measurement	48	17
Chemical engineering	27	8
Audio-visual technology	27	4
Computer technology	24	15
Materials, metallurgy	21	15
Electrical machinery, apparatus, energy	21	11
Pharmaceuticals	15	15
IT methods for management	14	6
Basic materials chemistry	13	20

Legend: DOD = Department of Defense; DOE = Department of Energy; IT = information technology; NASA = National Aeronautics and Space Administration; NIH = National Institutes of Health. WIPO=World Intellectual Property Organization.

Source: GAO analysis of DOD, DOE, NASA, and NIH licenses. | GAO-18-327.

Note: This includes both commercial and noncommercial licenses with patents in these technology fields. The number of licenses is a count of licenses that include one or more patents in a given technology field; licenses for multiple patents in different primary technology fields are counted as one license in each field.

[a] This includes U.S. patent numbers provided by the agencies as of fiscal year 2017. The totals do not include patent applications or foreign patents that may have been licensed.

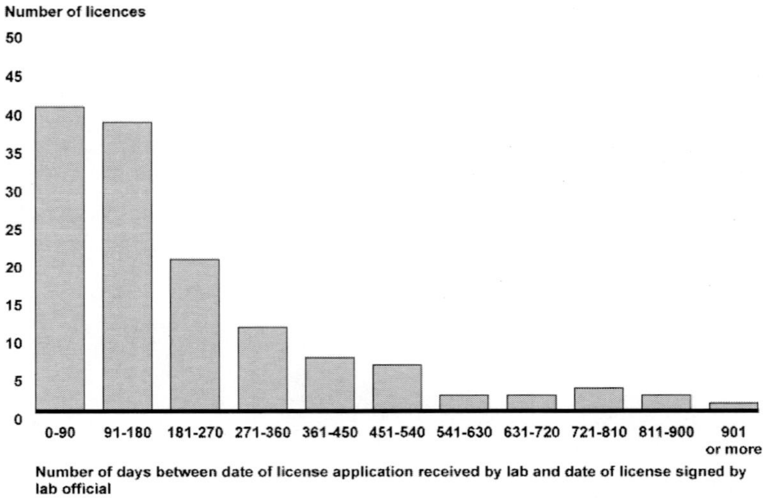

Source: GAO analysis of lisence agreements and application datees provided by Department of Defense (DOD), Department of Energy (DOE) National Aeronautics and Space Administration (NASA), and National institute of Health (NIH). | GAO-18-327.

Figure 4. Time from License Application Date to Signature of Commercial Patent Licenses Effective in Fiscal Year 2014 at DOD, DOE, NASA, and NIH.

Table 3. Use of Financial Terms in Commercial and Noncommercial Licenses Effective in Fiscal Year 2014 at DOD, DOE, NASA, and NIH

License type	No cost to licensee (percent)	Up-front fees (percent)	Minimum payments (percent)	Mile-stone payment (percent)	Equity (percent)	Earned royalties (percent)			
						Net sales	Gross sales	Other	Total
Commercial	na	84	84	24	5	68	17	7	89
Noncom-mercial	18	82	10	na	na	na	na	na	na

Legend: DOD = Department of Defense; DOE = Department of Energy; na = not applicable; NASA = National Aeronautics and Space Administration; NIH = National Institutes of Health.

Source: GAO analysis of DOD, DOE, NASA, and NIH licenses. | GAO-18-327.

Notes: Percentages of licenses incorporating various financial terms. Data for 6 of the 150 commercial licenses were not available. Milestone payments include any financial terms that depend on a defined event other than (1) the signing of the license, (2) a volume of sales of a product, or (3) a standard reporting period. Minimum payments include any type of recurring dollar payment that must be made in order to maintain the license, such as minimum annual royalties.

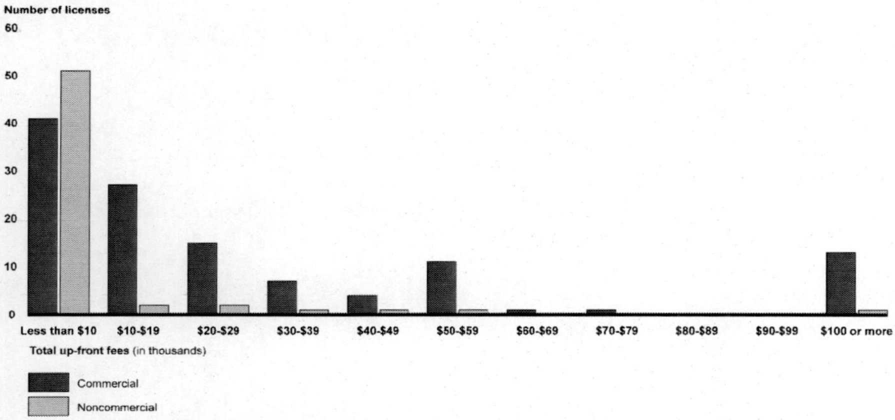

Source: GAO analysis of lisence agreements and application datees provided by Department of Defense (DOD), Department of Energy (DOE) National Aeronautics and Space Administration (NASA), and National institute of Health (NIH). | GAO-18-327.

Figure 5. Total Up-front Fees in Commercial and Noncommercial Licenses Effective in Fiscal Year 2014 at DOD, DOE, NASA, and NIH.

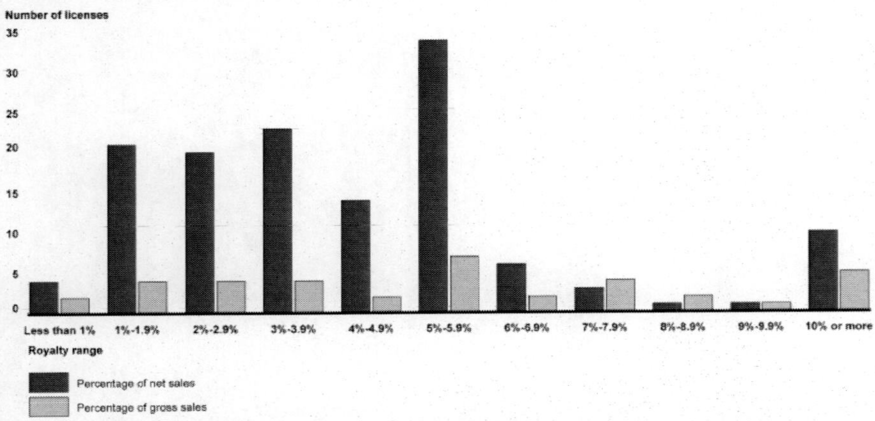

Source: GAO analysis of lisence agreements and application datees provided by Department of Defense (DOD), Department of Energy (DOE) National Aeronautics and Space Administration (NASA), and National institute of Health (NIH). | GAO-18-327.

Figure 6. Use of Specific Royalty Rates in Commercial Licenses Effective in Fiscal Year 2014 at DOD, DOE, NIH, and NASA (Royalties Based on a Percentage of Net or Gross Sales).

APPENDIX IV: COMMENTS FROM THE DEPARTMENT OF COMMERCE

UNITED STATES DEPARTMENT OF COMMERCE
The Secretary of Commerce
Washington, D.C. 20230

May 24, 2018

Mr. John Neumann
Director, Natural Resources and Environment
U.S. Government Accountability Office
441 G Street, NW
Washington, DC 20548

Dear Mr. Neumann:

Thank you for the opportunity to review and comment on the Government Accountability Office (GAO)'s draft report titled *Federal Research: Additional Actions Needed to Improve Licensing of Patented Laboratory Inventions* (GAO-18-327).

On behalf of the Department of Commerce, I have enclosed our comments on the draft report.

We concur with the recommendations and will take steps to implement them. If you have any questions, please contact MaryAnn Mausser, Department of Commerce Audit Liaison, at (202) 482-8120.

Sincerely,

Wilbur Ross

Enclosure

Department of Commerce's Comments on
GAO Draft Report titled *Federal Research: Additional Actions Needed to Improve Licensing of Patented Laboratory Inventions*
(GAO-18-327)

The Department of Commerce has reviewed the draft report, and we offer the following comments for GAO's consideration. Page numbers refer to page numbers in the report unless otherwise stated.

General Comments

Overall, it should be recognized that negotiating licenses is not a formulaic process and requires training and experience similar to any professional skill. Congress specifically directed Federal agencies to give priority to small businesses in the licensing process. Dealing with a small startup and a large multinational corporation are very different processes, for example. The degree of development, potential market, price, profit margin, exit strategy, required development funding, ability to attract investment, manufacturing cost and location, barriers to market entry, and a host of other factors affect the terms offered. Training and communication are very important, as noted in the report, and we support those principles.

Comments on Recommendations
The Government Accountability Office (GAO) made three recommendations to the Department of Commerce in the report.

- **Recommendation 1:** The Secretary of Commerce should instruct NIST to fully report the range of challenges in Federal patent licensing, such as those outlined in this report, by, for example, leveraging its survey of practices at Federal technology transfer offices, past studies conducted by the Federal Laboratory Consortium for Technology Transfer (FLC), and agency reports, and include that information in its summary reports to Congress.

  Commerce Response: The Department of Commerce concurs with this recommendation.

  NIST will work with the Interagency Workgroup for Technology Transfer to identify and provide further details and greater breadth regarding challenges in licensing in the annual report on Federal technology transfer. NIST will create a specific section to address this recommendation.

- **Recommendation 2:** The Secretary of Commerce should instruct NIST to clarify the link between the establishment of patent license financial terms and the goal of promoting commercial use, through appropriate means, such as the upcoming rulemaking process and updating relevant guidance.

  Commerce Response: The Department of Commerce concurs with this recommendation.

As noted, NIST is undertaking a review of all technology transfer processes of the Federal Government as part of the President's Management Agenda Cross Agency Priority Goal. This process includes an extensive comment period that NIST will closely analyze. This recommendation is noted for action by NIST. The final form of the actions to implement this recommendation, whether regulatory, policy, or otherwise, will be addressed in the outcome of the larger review process. As NIST is still in the data collection period, NIST has not yet determined the exact disposition of this matter.

- **Recommendation 3:** The Secretary of Commerce should instruct NIST to facilitate formal information sharing among the agencies to provide Federal labs with information on financial terms in comparable patent licenses, as appropriate.

  Commerce Response: The Department of Commerce concurs with this recommendation.

  NIST will work with the Interagency Workgroup for Technology Transfer and Federal Laboratory Consortium for Technology Transfer to examine and implement potential solutions. There is a great deal of concern regarding the security of a database with specific financial terms, as these terms are considered proprietary information of the company. Maintaining such a database would require a very high level of security and access control. Additionally, there are issues with information sharing across Government-operated and contractor-operated facilities with regards to business proprietary information from third-party sources. However, a range of potential solutions is possible, including identifying licensing officers in the various agencies with expertise to create a community of practice in various sectors and share best practices and approaches. NIST will also work with the aforementioned parties on any policy or guidance that can be incorporated in the FLC's Desk Reference, other training material, or other sources of information. As noted in response to recommendation 2, the larger review of Federal practices in this area is likely to influence the final form of the response to this recommendation.

APPENDIX V: COMMENTS FROM THE DEPARTMENT OF DEFENSE

RESEARCH AND ENGINEERING

OFFICE OF THE UNDER SECRETARY OF DEFENSE
3030 DEFENSE PENTAGON
WASHINGTON, DC 20301-3030

MAY 23 2018

Mr. John Neumann
Director, Natural Resources & Environment
U.S. Government Accountability Office
441 G Street, NW
Washington DC 20548

Dear Mr. Neumann:

This is the Department of Defense (DoD) response to the GAO Draft Report, GAO-18-327, 'FEDERAL RESEARCH: Additional Actions Needed to Improve Licensing of Patented Laboratory Inventions,' dated April 23, 2018 (GAO Code 100934).

The Department is providing official written concurrence with the one recommendation directed to DoD, Recommendation 4 without comments.

The primary action officer concerning this matter is Dr. Jagadeesh Pamulapati, whom you may reach at (571) 372-6372 or by email at jagadeesh.pamulapati.civ@mail.mil.

Mary J. Miller for
Performing the Duties of the
Assistant Secretary of Defense
for Research and Engineering

OFFICE OF THE UNDER SECRETARY OF DEFENSE
3030 DEFENSE PENTAGON
WASHINGTON, DC 20301-3030

RESEARCH
AND ENGINEERING

GAO DRAFT REPORT DATED APRIL 23, 2018
GAO-18-327 (GAO CODE 100934)

"FEDERAL RESEARCH: Additional Actions Needed to Improve Licensing of Patented Laboratory Inventions"

DEPARTMENT OF DEFENSE COMMENTS
TO THE GAO RECOMMENDATION

RECOMMENDATION 4: The Secretary of Defense should ensure that the agency or its labs document processes for establishing license financial terms, while maintaining flexibility to tailor the specific financial terms of each license. (Recommendation 4).

DoD RESPONSE: The DoD concurs with the recommendation. Upon issuance of the final GAO Report, DoD will prepare guidance for the Military Departments and appropriate Defense Agencies directing them to have their labs establish documentation of their licensing processes as appropriate.

Appendix VI: Comments from the Department of Energy

Department of Energy
Washington, DC 20585

May 23, 2018

Mr. John Neumann
Director
Natural Resources and Environment
U.S. Government Accountability Office
441 G Street, N.W.
Washington, D.C. 20548

Dear Mr. Neumann:

Thank you for providing a draft copy of the Government Accountability Office (GAO) report, "Federal Research: Additional Actions Needed to Improve Licensing of Patented Laboratory Inventions (GAO-18-327)." We have reviewed the draft report and provide the following comments below.

The Department intends to coordinate with the national labs to develop and document common practices in an attempt to make consistent some of the approaches to negotiating patent license financial terms. Although specific processes for establishing financial terms cannot be prescribed from the Department level, we will work with the labs to document minimum processes in order to mitigate some inconsistencies and ease inefficiencies across the national labs.

If you have any questions, please contact me, or Clara Asmail, clara.asmail@hq.doe.gov, 202-586-5471.

Sincerely,

Chanette Armstrong
Director, Office of Technology Transitions

Enclosure

Response to Report Recommendations

Recommendation: The Secretary of Energy should ensure that the agency or its labs document processes for establishing financial terms, while maintaining flexibility to tailor the specific financial terms of each license. (Recommendation 5)

Management Response: Concur

Negotiating terms for patent licensing is influenced by variety of different factors related to the condition of the technology and intended market as well as specific attributes of the parties. There are many complex and nuanced considerations that the professional licensing negotiator must always consider, including the: "readiness" of the technology and degree to which additional R&D or marketing investments may be needed; risk involved in those investments compared to incumbent and competing solutions; size of the market and its elasticity to adopting new technologies; license exclusivity, or non-exclusivity, and field of use; availability of inventor for continued co-development; and resources at the potential licensee to include financial, business, technical and marketing strengths. Especially for licensing negotiators of technologies developed at federal labs these factors are weighed against urgencies and alternate pathways to move the technology toward practical application. Because of the many and nuanced dimensions that impact license negotiations, it would be impractical to develop a prescriptive approach to establish patent license financial terms.

Several statutes including Bayh-Dole Act, the Stevenson-Wydler Act, the Federal Technology Transfer Act, and the National Competitiveness Technology Transfer Act of 1989 (NCTTA), authorize the contractor-operated DOE laboratories to retain title to inventions created using federal funding, pursue patent protection, and license those inventions to third parties. While DOE establishes certain requirements for its lab contractors related to licensing DOE technologies, those requirements relate to non-financial considerations such as U.S. Competitiveness and successor contractor issues.[1] The financial portions of license agreements for national laboratory technologies are between the lab contractor and the licensing entity (licensee).

Nevertheless, the Department of Energy agrees that some standardization of practices by documenting minimum processes can mitigate some inconsistencies and ease efficiencies across the national labs. The DOE intends to encourage the labs that when practical with regard to time and funding resources, to make use of market analyses and benchmarking tools and services that are commercially available. DOE will coordinate with the labs to collect documentation and develop guidance that can guide the labs as they negotiate patent licenses especially for effective information gathering and basic guidelines for establishing and leveraging communities of practice. The documentation of internal best practices to be developed will complement the Licensing Guide and Sample License which was developed by DOE labs responsible for licensing patents and intended to raise awareness among potential licensees about general DOE patent license terms to be expected.

The estimated completion date for this activity is July 30, 2019.

[1] *See* 48 C.F.R. §970.5227-3.

APPENDIX VII: COMMENTS FROM THE DEPARTMENT OF HEALTH AND HUMAN SERVICES

DEPARTMENT OF HEALTH & HUMAN SERVICES   OFFICE OF THE SECRETARY

Assistant Secretary for Legislation
Washington, DC 20201

MAY 2 4 2018

John Neumann
Director, Natural Resources and Environment
U.S. Government Accountability Office
441 G Street NW
Washington, DC 20548

Dear Mr. Neumann:

Attached are comments on the U.S. Government Accountability Office's (GAO) report entitled, "*Federal Research: Additional Actions Needed to Improve Licensing of Patented Laboratory Inventions*" (GAO-18-327).

The Department appreciates the opportunity to review this report prior to publication.

Sincerely,

Matthew D. Bassett
Assistant Secretary for Legislation

Attachment

GENERAL COMMENTS FROM THE DEPARTMENT OF HEALTH & HUMAN SERVICES ON THE GOVERNMENT ACCOUNTABILITY OFFICE'S DRAFT REPORT ENTITLED - FEDERAL RESEARCH: ADDITIONAL ACTIONS NEEDED TO IMPROVE LICENSING OF PATENTED LABORATORY INVENTIONS (GAO-18-327)

The U.S. Department of Health & Human Services (HHS) appreciates the opportunity from the Government Accountability Office (GAO) to review and comment on this draft report.

GAO Recommendation 7:
The Director of the National Institutes of Health (NIH) should ensure that the agency or its labs document processes for establishing license financial terms, while maintaining flexibility to tailor the specific financial terms of each license.

HHS Response:
HHS concurs with GAO's recommendation.

The report notes four potential aspects of the process for establishing financial terms:

1. Factors to consider
2. Sources of information
3. Methods for developing financial terms
4. Use of specific types of financial terms

NIH will develop materials to include all four aspects of the process and provide materials for technology transfer staff by October 1, 2018. NIH will share its materials with the Interagency Working Group for Technology Transfer.

The financial terms of licenses typically include upfront fees, minimum annual payments, running royalties as a percentage of sales, milestone payments, a percentage of sublicensing revenue, and a percentage of the cash value of an acquisition of a small company or a unit focusing primarily on the licensed technology.

NIH already conducts one-on-one and group training for licensing specialists. NIH will develop and post for internal use more detailed information than what is covered in training sessions. The materials will include the requirements for a licensee to provide detailed commercialization plans/development timelines with their license application and descriptions of the use of the types of royalties that are negotiated. The materials will provide an explanation of how these aspects of licensing focus licensees' efforts toward translating an invention into a product to support public health needs.

Training materials will also convey complexity and nuance of negotiating terms for commercial patent licenses. The terms are based on variables such as the type of technology; the development stage of the technology (usually very early); the breadth of the patent claims and its status (pending or issued, United States and foreign); the market size and risk involved in developing the product; potential degrees of exclusivity and fields of use of the technology; the size of the company and its needs; and the availability of the government's inventor to assist with questions and testing. A company may agree to take on greater responsibility to address higher-risk or less lucrative markets to benefit the public but may ask for lower royalty rates in return. Ultimately, negotiations between NIH and the company include these aspects.

Page 1 of 1

Appendix VIII: Comments from the National Aeronautics and Space Administration

National Aeronautics and Space Administration

Office of the Administrator
Washington, DC 20546-0001

May 24, 2018

Mr. John Neumann
Director
Natural Resources and Environment
United States Government Accountability Office
Washington, DC 20548

Dear Mr. Neumann:

The National Aeronautics and Space Administration (NASA) appreciates the opportunity to review and comment on the Government Accountability Office (GAO) draft report entitled, "Federal Research: Additional Actions Needed to Improve Licensing of Patented Laboratory Inventions" (GAO-18-32) dated April 23, 2018.

In the draft report, GAO makes one recommendation to the NASA Administrator intended to improve license financial terms for its labs while maintaining flexibility to tailor the financial terms of each license.

Specifically, GAO recommends the following:

Recommendation 1: The Administrator of NASA should ensure that the Agency or its labs document processes for establishing license financial terms, while maintaining flexibility to tailor the financial terms of each license.

Management's Response: Partially concur. Setting financial terms for commercial patent licenses is complicated and nuanced. Just as each technology, application, and market is different, so are the financial terms of each license. Relevant factors include the development stage of the technology; potential degrees of exclusivity; whether the Government has an inventor available to assist with questions and testing; the field of use; the length of time it will take to bring the product to market; resources available for market analysis; the skill and experience of the licensing agent; and many other considerations. Further complicating this process, setting financial terms of licenses is usually an iterative process between the Government and potential licensees. It is a negotiation.

It is, therefore, not surprising that none of the agencies examined under this audit have documented established guidelines for establishing financial terms of patent licenses. Nonetheless, NASA has, as the audit report notes, published considerations and parameters as well as stated intentionality in setting terms.

While the parameters we consider for financial terms are established, the methods and processes for gathering this information vary depending on available resources. Nevertheless, while it may not be possible to predict available resources or clearly predict the financial terms of licenses without knowing the parameters of each deal, NASA agrees that it can make efforts to standardize, document, and communicate the minimum process by which it sets financial terms of patent licenses.

We are convening the NASA licensing managers this summer for a meeting, and we will use that time to discuss standardization of documentation of processes. The goal will be to produce two documents, one an internal set of instructions for setting financial terms and the other a public-facing set of guidelines for how NASA sets terms.

Estimated Completion Date: NASA will have draft guidelines for setting financial terms of patent licenses completed by June 15, 2018. This draft will be shared with Headquarters and Field Centers' technology transfer management and the NASA Office of General Counsel. A final version of each will be completed by August 15, 2018, at which time the internal standards will be shared internally with all Field Centers' technology transfer staff. NASA will post the external-facing document on its technology licensing Web site by September 30, 2018. NASA will also share its final products with the Interagency Working Group for Technology Transfer by October 31, 2018.

Once again, thank you for the opportunity to review and comment on the subject draft report. If you have any questions or require additional information, please contact Daniel Lockney (Program Executive for Technology Transfer) on (202) 358-2037, or G. Michael Green (Space Technology Mission Directorate Chief of Staff) on (202) 358-4710.

Sincerely,

Stephen G. Jurczyk
Associate Administrator (Acting)

GAO CONTACT AND STAFF ACKNOWLEDGMENTS

GAO Contact

John Neumann, (202) 512-3841 or neumannj@gao.gov

Staff Acknowledgments

In addition to the contact named above, Robert J. Marek (Assistant Director), James D. Ashley, Kevin S. Bray, Virginia A. Chanley, Ellen L. Fried, Sarah C. Gilliland, Cheryl M. Harris, Robert Letzler, Gregory A. Marchand, Christopher P. Murray, Emmy L. Rhine Paule, Dan C. Royer, Ardith A. Spence, Vasiliki Theodoropoulos, and Reed Van Beveren made key contributions to this report.

REFERENCES

Bozeman, Barry. *Technology Transfer Research and Evaluation: Implications for Federal Laboratory Practice*, Final Report to VNS Group, Inc. and the U.S. National Institute of Standards, 2013. Accessed on March 14, 2018. https://www.nist.gov/tpo/return-investment-roi-initiative.

Bozeman, Barry, Heather Rimes, and Jan Youtie. "The Evolving State-of-the-Art in Technology Transfer Research: Revisiting the Contingent Effectiveness Model." *Research Policy*, vol. 44, no. 1 (2014): 34-49.

Franza, Richard M. and Kevin P. Grant. "Improving Federal to Private Sector Technology Transfer: A Study Identifies Seven Critical Factors with the Greatest Impact on Whether Transfer Attempt Succeeds or Fails." *Research Technology Management*, vol. 49, no. 3 (2006): 36-40.

Greiner, Michael A. and Richard M. Franza. “Barriers and Bridges for Successful Environmental Technology Transfer.” *Journal of Technology Transfer*, vol. 28, no. 2 (2003): 167-177.

Howieson, Susannah V., Stephanie Shipp, Gina Walejko, Pamela Rambow, Vanessa Peña, Sherrica S. Holloman, and Phillip N, Miller. *Exemplar Practices for Department of Defense Technology Transfer*. Alexandria, Va.: Institute of Defense Analyses, January 2013.

Hughes, Mary E., Susannah V. Howieson, Gina Walejko, Nayanee.

Gupta, Seth Jonas, Ashley T. Brenner, Dawn Holmes, Edward Shyu, and Stephanie Shipp. *Technology Transfer and Commercialization Landscape in the Federal Laboratories*. Alexandria, Va.: Institute of Defense Analyses, June 2011.

Jin, D., X. Mo, A. M. Subramanian, K. H. Chai, and C. C. Hang. “Key Management Processes to Technology Transfer Success.” *2016 IEEE International Conference on Management of Innovation and Technology*, (2016), 67-71.

Linton, Jonathan D., Cesar A. Lombana, and A. D. Romig, Jr. “Accelerating Technology Transfer from Federal Laboratories to the Private Sector—the Business Development Wheel.” *Engineering Management Journal*, vol. 13, no. 3 (2001): 15-20.

Office of Science and Technology Policy and the National Institutes of Health, National Heart, Lung and Blood Institute. *Lab-to-Market Inter-agency Summit: Recommendations from the National Expert Panel*.

Washington, D.C.: *National Expert Panel*, White House Conference Center, May 2013.

Stepp, Matthew, Sean Pool, Nick Loris, and Jack Spencer. *Turning the Page: Reimagining the National Labs in the 21st Century Innovation Economy*. Washington, D.C.: Information Technology and Innovation Foundation, Center for American Progress, and Heritage Foundation, June 2013.

Toregas, Costis, E. Colin Campbell, Sharon S. Dawes, Harold B. Finger, Michael D. Griffin, and Thomas Stackhouse. *Technology Transfer: Bringing Innovation to NASA and the Nation*. Washington, D.C.: National Academy of Public Administration, November 2004.

U.S. Department of Energy, Commission to Review the Effectiveness of the National Energy Laboratories. *Securing America's Future: Realizing Potential of the Department of Energy's National Laboratories*, vol. 1, Executive Report. Washington, D.C.: October 2015. Accessed March 14, 2018. https://www.energy.gov/labcommission/downloads/final-reportcommission-review-effectiveness-national-energy-laboratories.

Wang, Mark, Shari Pfleeger, David M. Adamson, Gabrielle Bloom, William Butz, Donna Fossum, Mihal Gross, et al. *Technology Transfer of Federally Funded R&D: Perspectives from a Forum. Conference Proceedings*. Santa Monica, Calif.: RAND Corporation, 2003.

Obtaining Copies of GAO Reports and Testimony

The fastest and easiest way to obtain copies of GAO documents at no cost is through GAO's website (https://www.gao.gov). Each weekday afternoon, GAO posts on its website newly released reports, testimony, and correspondence. To have GAO e-mail you a list of newly posted products, go to https://www.gao.gov and select "E-mail Updates."

Order by Phone

The price of each GAO publication reflects GAO's actual cost of production and distribution and depends on the number of pages in the publication and whether the publication is printed in color or black and white. Pricing and ordering information is posted on GAO's website, https://www.gao.gov/ordering.htm.

Place orders by calling (202) 512-6000, toll free (866) 801-7077, or TDD (202) 512-2537.

Orders may be paid for using American Express, Discover Card, MasterCard, Visa, check, or money order. Call for additional information.

Connect with GAO

Connect with GAO on Facebook, Flickr, Twitter, and YouTube. Subscribe to our RSS Feeds or E-mail Updates. Listen to our Podcasts. Visit GAO on the web at https://www.gao.gov.

To Report Fraud, Waste, and Abuse in Federal Programs

Contact:
Website: https://www.gao.gov/fraudnet/fraudnet.htm
Automated answering system: (800) 424-5454 or (202) 512-7470.

Congressional Relations

Orice Williams Brown, Managing Director, WilliamsO@gao.gov, (202) 512-4400, U.S. Government Accountability Office, 441 G Street NW, Room 7125, Washington, DC 20548.

Public Affairs

Chuck Young, Managing Director, youngc1@gao.gov, (202) 512-4800 U.S. Government Accountability Office, 441 G Street NW, Room 7149 Washington, DC 20548.

STRATEGIC PLANNING AND EXTERNAL LIAISON

James-Christian Blockwood, Managing Director, spel@gao.gov, (202) 512-4707 U.S. Government Accountability Office, 441 G Street NW, Room 7814, Washington, DC 20548.

INDEX

E

F

G

H

I

J

L

M

N

O

P

Q

R

S

T

U

V

W

Y